The Birth of China Seen through Poetry

About the Author

Though a devoted lover of poetry and history all his life, the author is by profession a scientist specialized in theoretical particle physics. Thus, his poetry, gathered at odd moments over a lifetime, is to him but an amateur's private affair, and only the hope now of helping to promote inter-cultural understanding gives him the incentive and daring to publish some of it. Born in Guangzhou, China, to Chinese parents, brought up mostly in Hong Kong, then having lived and worked since in various places around the world, he has naturally a particular interest in how cultures mix.

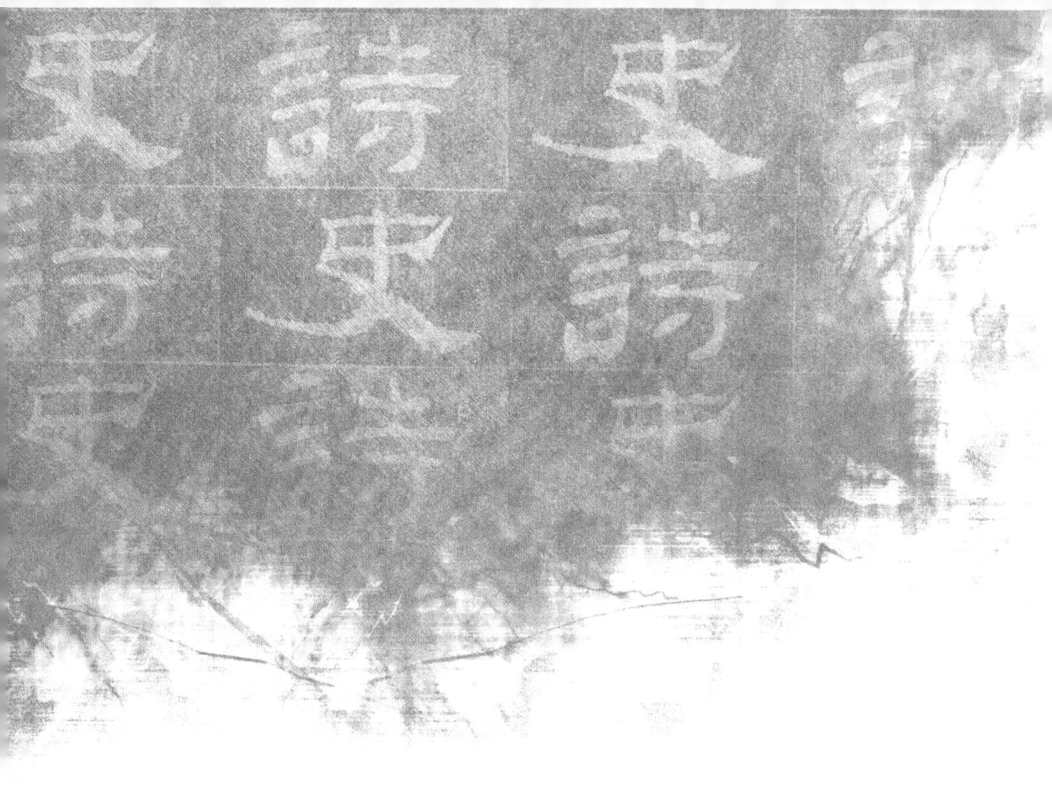

The Birth of CHINA Seen through Poetry

Chan Hong-Mo

World Scientific

NEW JERSEY · LONDON · SINGAPORE · BEIJING · SHANGHAI · HONG KONG · TAIPEI · CHENNAI

Published by

World Scientific Publishing Co. Pte. Ltd.

5 Toh Tuck Link, Singapore 596224

USA office: 27 Warren Street, Suite 401-402, Hackensack, NJ 07601

UK office: 57 Shelton Street, Covent Garden, London WC2H 9HE

British Library Cataloguing-in-Publication Data
A catalogue record for this book is available from the British Library.

THE BIRTH OF CHINA SEEN THROUGH POETRY

ISBN-13 978-981-4335-33-1 (pbk)
ISBN-10 981-4335-33-9 (pbk)

Email: enquiries@stallionpress.com

Printed in Singapore.

To the memory

of

my father who taught me since early childhood to appreciate poetry,

and

my mother who first showed me by her love what it all meant.

Contents

Introduction

Much of the conflict in this world is due to misunderstanding. People from another culture seem strange to us not only because they speak a different language but also because they think different thoughts. Their language we can learn given time, but to understand the way they think, we shall have to dig back into their history to trace the origin of their thought pattern. And for most of us, it would be no easy matter to learn enough of the past of another culture to undo the layers of wrapping imposed on it by history and recognize the common humanity hidden underneath.

At this time when China is beginning to take up a position in the world community more commensurate to its population and size and hence to pull its weight in shaping our common future, many would no doubt wish to know more about its ancient culture. To most English readers, however, ancient Chinese history is probably a closed book, first because there is so much of it, and second because it may seem such a bore. Indeed, why should it bother us to know that, say, some few thousand years ago, an emperor named so-and-so lost a battle at a God-forsaken spot somewhere in China and lost thereby his throne? The trouble is that most of recorded history is a chronicle of the actions of the great, that is, the great of their days. But those days have long since vanished and their actions are now of little consequence. Nevertheless, those happenings also affected the lives of the people living at that time and moulded their thoughts, and it is the totality of these thoughts which has come down to us today as culture. Of these, unfortunately, recorded history usually tells us rather little, not being quite the medium for the job, and even what little it does say is cast in such terms as would need the interpretation of an expert historian.

There is however another window through which the ancients can speak directly to us. After all, they were just humans as we are, and as we do, they sang when they were happy and moaned when they were sad. And some of their songs and moanings have crystallized and come down to us as poetry. Through poetry then, barring the obstacle of language, most of us can have a peep at ancient China, and further, if it is placed in the proper historical context, reconstruct for ourselves how Chinese culture developed into the form we see today. It is with this purpose in mind that the idea for the present little volume is conceived. And in this, one is helped by the Chinese nation's particular fondness for both history and poetry which has left us with an unbroken historical record and poetic tradition over nearly three thousand years.

The author freely confesses that no altruistic ideal for the promotion of intercultural understanding, however worthwhile that may be, would have prompted him to attempt the thankless and bound-to-failure task of translating poetry, especially between languages as different as Chinese and English, were it not for his love for the poetry itself. Actually for him, the logic has gone the other way round. Being fond of poetry and of making up verses himself, but having often neither the inspiration nor the deep emotional experience needed for original creations, he has sought a degree of satisfaction by translating the work of others, and then, having done so, made the promotion of intercultural understanding an excuse for doing it. Nevertheless, it can be hoped that the excuse is not an entirely lame one, and that some readers may indeed understand Chinese culture more from this work, and perhaps even catch a glimpse of the beauty of the Chinese originals from the translated poems.

The period covered starts from the beginning of continuous recorded Chinese history around the eighth century B.C. and ends shortly after the fall of the Northen Song dynasty in the twelfth century A.D., with just a sprinkling of poems in the epilogue to represent modern times. Though determined largely by the author's own limited knowledge and taste, the selection reflects perhaps also a qualitative change in society's attitude to poetry, possibly even in society itself, at the end of that period, when the development of Chinese poetry in its classical form slowed considerably from its erstwhile heady pace.

Before proceeding to the actual material, a few observations on the Chinese language would be in order. There is a certain amount of inconsistency in common usage between the terms China, the country and Chinese, the language, much as that existing between, say, England and English. China as a country comprises some fifty odd ethnic groups with different cultural and linguistic backgrounds, while Chinese refers to the language which, though spoken all over the country, is the mother tongue of only the dominant Han majority which accounts for some ninety-five percent of the total population. With due apology to members of the other ethnic groups, the author will perpetuate this inconsistency by citing in this work only poems written in Han Chinese, which is unfortunately unavoidable since he does not know any of the other languages. The ethnic minorities are represented, and therefore inadequately, only by those of their poets who happened to have written in the Han language, of which however there are extant some very powerful examples. One should add also that what one calls the Han ethnic group is itself but an amalgam of numerous cultures merged together in ancient times, both prehistoric and historic, only so coherently as to appear to us now near uniform.

The Chinese (Han) language belongs to the so-called Sino-Tibetan family the speakers of which are second in number only to those of the Indo-European family in today's world. Like its close cousins such as Vietnamese, Han Chinese is distinguished by having tones. The characters are all monosyllabic but are each assigned a "tone" (*sheng*) out of a list of from four to nine or ten, depending on the dialect. A *sheng* is actually not a pure note but rather a sequence of notes on the musical scale. In modern Cantonese, the author's mother tongue, which is thought to approximate to some extent the ancient dialect in which most of the cited poems were written, there are eight main *sheng*, divided into two groups of four each, the two groups being separated by somewhat less than an octave. The four *sheng* in each group, called *ping, shang, qu,* and *ru*, are represented respectively by a roughly level, a rising, a descending then rising, and a descending sequence of notes on the musical scale. The absolute scale does not matter; so each person speaks in a key best suited to themselves.

The tonal structure may seem complicated and is indeed not easy to master for a non-Chinese speaker. But even without mastering it, one could imagine that a Chinese poem which fully exploits this tonal quality is almost like a song when read aloud. To hear examples of what it actually sounds like, see Appendix III. Indeed, one can say that the tonal structure gives to the Chinese poet an extra dimension to explore in addition to that of rhythm open to, say, the English poet, and thus to the Chinese poem an additional appeal. A drawback, perhaps, is that the added attention needed for tonal quality inhibits the poet from creations of great lengths. There is thus no Chinese equivalent to the epic poem in Indo-European tongues. Also, it might be argued that Chinese music suffered as a result of too close an association with poetry because of the inherent musical quality of the latter, an association which deterred Chinese music from developing freely on its own and attaining the same heights as in some other advanced cultures.

In English translation, of course, the tonal quality of the original is lost, although attempts have been made in most examples here to keep the original structure (as regards line lengths and rhyming patterns, for example) and flow. Despite this inherent shortcoming, however, the hope is that, poetry being universal, it will still find resonance within our breast, could we but understand what is being said. And this is just what the present author attempts to do, bringing, not so much the poems as the poetry, across the wide separation in both time and space from ancient China to his readers.

Finally, a few words on the special characteristics of this little volume. Though comprising some eighty poems in different moods and styles scattered over nearly three thousand years, it is constructed as a single unit. Running through the book mostly on left-hand pages is a narrative outlining very briefly the history and poetic tradition of China, while poems from the corresponding periods are printed on right-hand pages. It can thus be read straight through with the poems serving as illustrations for the narrative and the narrative as a backdrop to the poems. The poems themselves, though rendered into English by an amateur with Chinese as first language, can at least claim to have been crafted with loving care and a deep regard for the original. Besides, though selected and rendered by the author just as the mood hit him, that this has been done over a long time ensures a degree of randomness in the process and makes the collection reasonably representative. It may thus serve the dual purpose of a small anthology to be kept, returned to at leisure, and enjoyed.

To help the reader place the poems selected and the events mentioned in the narrative, a time-line with maps is given in Appendix I. And for readers who may be interested, the text of the original poems is reproduced in Appendix II and a recorded recital of the poems will be made available on the web. (For details see Appendix III.) A glossary of Chinese names and terms is provided in Appendix V. The historical narrative presented is the result of a lifetime of reading for pleasure, the sources for which would be difficult properly to trace, but I give in Appendix II a short list of the sources from which I think I have learnt the most. Of some of the poems, several versions are known with small variations and some, of course, are given different interpretations by different experts, in which case, I have just selected the version and interpretation I appreciate the most. I have, however, deliberately and meticulously avoided consulting other translations of these poems so as to ensure that nothing should stand between the Chinese original and my readers, except of course, unavoidably, myself. In Appendix IV are captions of the illustrations, in case some readers may wish to know what they represent. It may help my readers first to have a quick glance at these appendices so that they know what information is available there before starting on the main text. This will save breaking the narative frequently for cross references.

In the transliteration of Chinese names both in the poems and in the narrative, I have followed the official *pinyin* system. For the accurate phonetic values of the various symbols, the reader has to be referred elsewhere. I shall only point out the following for the casual reader. The vowels: *a, e, i, o, u* are pronounced roughly as: *ah, er, ee, oh, oo* in English. Of the consonants, I shall single out only the, to English readers, disconcerting *q* which (whether followed by *u* or not) is pronounced rather like *ts*.

That this little volume has taken the form it does, or even that it was attempted at all, owes much to my family and friends who have each made an imprint on my life, and through me on to this little book. Among those who have contributed directly to this work, I wish first to thank my test-readers who have kindly read through a draft version and given me encouragement and valuable suggestions leading to material improvements: Prof. Jose Bordes, my daughter Man-Suen (Chan, Dr.), Cllr. Janet Morgan, Dr. Ali Namazie, Prof. Sir David Todd, and Dr. Giulio Villani. Some others have given me assistance and advice, such as my daughter Man-Kwun (Chan) and her husband Barnabas Baggs, Mr. David Carlson and Ms Judy Tsou. Special thanks are due to Prof. Evangeline Almberg-Ng and my brothers Hong-Ching (Chan, Eric) and Hong-Fat (Chan, Henry). Apart from serving as test-readers, Evangeline has used her professional expertise in translation to help polish my amateurish efforts, while my brothers have contributed, the one his special calligraphy and the other his carved seals, to embellish the text. And lastly, to Prof. Yau Shun-chiu and my wife Sheung Tsun (Tsou, Dr.) I am indebted for more than common help. Yau-yau, as we call him, has spent much of his valuable time, not only in applying his knowledge, considerable in both depth and breadth in all things Chinese, to improve my work, both on broad issues such as the balance of material and in details such as the accuracy of historical facts, but also in seeking and checking for me some information and typing for me some Chinese texts. Sheung Tsun, on the other hand, apart from general advice and encouragement and several careful readings of the manuscript, has made herself available at all times of the day throughout the many years that this work was in preparation, to answer questions of all sorts, from poetic style and sentiment to punctuation and spelling. It is only because of my habitual reliance on their support in many things — and in the case of Sheung Tsun, this means essentially everything — I do, that their help is no longer consciously acknowledged, and they are thus able to escape being roped in as co-authors.

Prehistory

In the last half-century, Chinese archaeology has made great strides, from which a fairly clear picture of prehistoric China has now emerged.

By about 5000 B.C., China was in the late neolithic stage, with settlements scattered over the two great river basins, of the Huanghe (Yellow River) in the north, and of the Changjiang (Yangtze) in the south. Agriculture which had made its appearance already some thousands of years before, was by then widespread, with mainly millet cultivated in the drier north, but rice in the south. Pottery also was common, from which archaeolgists have identified distinct cultures coexisting in different regions.

Bronze artifacts started to make an appearance around 3000 B.C., though only in small quantities then. At the same time, fortified settlements or "cities" appeared all over the two great river basins. Some are quite large, covering tens of hectares, with one in the midst of the Changjiang basin covering as much as a square kilometre. The defensive walls of hardened earth alone, as judged by the remains, would have required some tens of thousands of dedicated man-years to build, which points to the existence of sizeable organised societies capable of mobilizing and maintaining such a labour force. Some archaeologists now think that this period could correspond to that of the "Five Emperors" (*Wu Di*) described in some fragmentary records formerly thought to be just legendary.

This situation lasted for about a thousand years with parallel development in various parts of the country. However, at around the end of the millennium, a sudden flourishing of the culture in the central Huanghe basin allowed it to supersede all others, and then gradually to spread, absorbing other cultures as it did so, giving birth eventually to a near uniform Chinese civilization.

The Spring and Autumn Period

The systematic recording of historical events in China which continued unbroken till the present day began at the end of the eighth century B.C. in the so-called Spring and Autumn period of the Eastern Zhou dynasty, from which date also the oldest poems cited below. The early records were called *chunqiu* (meaning spring and autumn) presumably because agriculture, the main occupation then, began in the spring and ended in the autumn, only after which could any major exploits such as military campaigns be undertaken, and the period was named after them. Lasting some three hundred years, this period saw a rapid development of Chinese culture in every direction, accompanied by a big increase in the population and in the land area occupied, both probably connected with the advent of iron tools around that time. Learning also flourished; both Kongzi (Confucius) and Laozi (Lao Tzu) were of that period.

The Zhou dynasty itself started in the eleventh century B.C. and lasted nearly eight hundred years. It was centred at first in the west in present-day Shaanxi but was forced by nomadic invaders to move its capital eastwards across the mountains to Luoyang in 770 B.C., after which it is known as the Eastern Zhou dynasty. Before Zhou, there had been in China two other historical dynasties, Shang (sixteenth to eleventh century B.C.) and Xia (twenty-first to sixteenth century B.C.), of which only fragmentary historical records remain, some handed down by tradition and some gleaned by archaeologists from inscriptions on bronze vessels and oracle bones. All three were Han ethnic dynasties centred in the Huanghe (Yellow River) basin, which is thus often regarded as the cradle of Chinese civilization.

This does not mean, however, that all Chinese civilization started there and had a Han ethnic origin. Major archaeological discoveries in recent decades have revealed that there were in prehistoric times a wealth of different neolithic and bronze age cultures all over China, not in any way inferior to their contemporaries in the Huanghe basin. For instance, the Liangzhu culture found in the lower reaches of the Changjiang (Yangtze) around present-day Hangzhou and dating from c. 3000 - 2000 B.C. has left us some beautiful examples of finely worked jade with few contemporary equals. Even as late as the end of the second millennium B.C. there was still in present-day Sichuan a culture advanced enough to maintain a capital city that rivalled in size, and to produce bronze artifacts that rivalled in sophistication their famed Shang dynasty equivalents, and yet had quite distinct characteristics. Indeed, the bronze figures discovered at Sanxingdui had features more akin to those found in the Middle East than in China today. These facts seem to indicate that Chinese culture in prehistoric and early historic times was multipolar, almost kaleidoscopic. But by the foundation of the Zhou dynasty around 1000 B.C., Han civilization appeared already to have established its dominance.

The Zhou king ruled as overlord with, at least nominal, vassals scattered over the Huanghe basin all the way eastwards to the sea, and some few as far flung as in modern Hunan, Zhejiang and Sichuan. At first, Zhou influence seemed limited to the vassal cities and their immediate environs, with vast tracts of land in between still open for nomads to hunt and wander in. However, as the population increased the city states expanded the area under their control, gradually squeezing the non-Han ethnic groups to the periphery. By the time of the Spring and Autumn period when our story begins, the "Middle Kingdom" had become already a seemingly solid mass of Han culture, though still with plenty of variations due no doubt to the many different local cultures it had absorbed.

Shijing is the oldest extant anthology of Chinese poetry dating from about the eleventh to the sixth century B.C. Tradition has it that the Zhou king, mindful of the inner feelings often expressed in songs, had assigned special envoys to tour the country collecting the ditties they heard so as to gauge the people's mood, the better to further their wishes and address their grievances. Whether it was really so collected or its purpose so altruistic, the result is a rich legacy of some three hundred poems collected from all corners of the kingdom which at that time covered an area comprising roughly the provinces of present-day China along the Huanghe (Yellow River) basin. (See map in Appendix I.) The most interesting to us now, perhaps, are the folk lyrics which though coming from a world very different from ours, still resonate with the same humanity that activates us, and move us with their simple dignity and grace. We suffer as they suffered, love as they loved, except that in their simpler world, their feelings appear more pristine and pure.

The society Shijing describes was feudal, with the common people mostly working as serfs on the land. The first poem cited, *Seventh Month*, coming from present-day Shaanxi and dating from the eighth century B.C. or before, gives a very vivid picture of how life was lived, although there were likely to be big variations over both time and space. The poem describes the everyday life of a peasant through the year. There were then two calendars in use, the older Xia (lunar) calendar, some version of which is still in use today for traditional festivals, and the Zhou calendar. In both, the months are labelled from one to twelve, but with the labels in the latter shifted forwards by two months. The first Zhou month corresponds roughly in time to January in the Gregorian calendar. The Seventh Month in the Xia calendar corresponds then roughly to September.

In the Seventh Month

From Shijing

In the Seventh Month, the heat abates;
In the Ninth Month, clothes are handed out;
In days of January, the cold wind starts.
The days of February are shiverng days:
Without a cloak or even a shirt for cover,
How are we to last the winter out?
In days of March, soil-breaking tools are mended.
When April comes, we peasants make a move,
Together with our wives and children all.
We take our meals at work on southern fields.
At least the foreman's happy with our chores.

In the Seventh Month, the heat abates;
In the Ninth Month, clothes are handed out.
In days of Spring, the sun begins to shine,
And the nightingales begin to sing,
While girls with giant baskets at their back
Move slowly along the narrow lanes
To pick the tender leaves of mulberry.
As days in Spring grow longer yet and longer,
They gather straws for silkworms more and more;
But the girls are worried sick at heart,
For fear the lords should take them home with them.

In the Seventh Month, the heat abates;
In the Eighth, we gather reeds in store.
The Silkworms' Month, we prune the mulberry bush:
Wielding axe or knife of any kind.
We cut the high-flung straggling branches off,
While propping up the weaker, younger trees.
In the Seventh Month, the thrushes call;
In the Eighth, the weaving's to be done.
Some cloth we dye in yellow, some dye black,
Some other still we dye a deep bright red,
For making undergarments for the lords.

The Fourth Month, medicinal fruits are picked,
While cigalas sing the Fifth Month through.
In the Eight Month, grain is harvested;
By the Tenth Month, leaves have fallen off.
January is spent in setting snares
To trap the fox and others of its kind
For furs to make our lords their winter coats.
In February we gather for the hunt,
And we're trained on how to handle arms.
The yearling catch we keep as private lots;
The bigger game are presents to Milord.

The Fifth Month, locusts loudly shake their legs;
The Sixth, grasshoppers chime in with their wings.
The Seventh, crickets stay out in the fields;
The Eighth, they come to shelter on our roof.
The Ninth, they enter at the cottage door;
The Tenth, we find them beneath our beds.
Then stop the gaps with daub, smoke out the rats —
Block the northern window, mend the gate,

For that, poor wife, and you my poor, poor child,
Is the passing of another year.
We're back in here to see the winter out.

The Sixth Month, they have various sorts of plums;
The Seventh, they cook haricot with greens.
The Eighth Month, they beat down the jujubes sweet;
The Tenth Month, crops of rice are harvested.
With these they brew the wine to drink in Spring,
To wish each other long and happy lives.
The Seventh Month, there are for us but marrows;
In the Eighth, we cut ourselves some gourds.
In the Ninth, we have for food but seeds of flax
And bitter greens. For fuel, but rotted posts.
That's how they feed us peasants the whole year through.

The Ninth Month, barns and threshing fields are built;
The Tenth, we turn in produce of the year.
There's millet, both the late and early crops,
And various grain, and flax, and beans and wheat.
But alas for peasants such as us,
Handing in our crops has not yet seen us through.
Inside the Manor we've still many chores:
The daytime, we cut cogongrass for thatch,
The evenings are spent in plaiting ropes.
Then quick, let's climb above to mend the roof;
It's time already sowing should begin.

In February, we dig out chunks of ice;
In March, we put them underground in stores.
And then one April morning, Spring returns,
And offering is made of lamb and chives.

But by the Ninth Month, frosty days are back;
The Tenth, the threshing field is cleaned and swept .
Then bottled wine is taken out from stores,
And lambs are slaughtered ready for the feast.
Together then we enter Manor Hall,
And holding big wine-cups of wild-ox horns,
We sing, "Long live, my lord, ten thousand springs!"

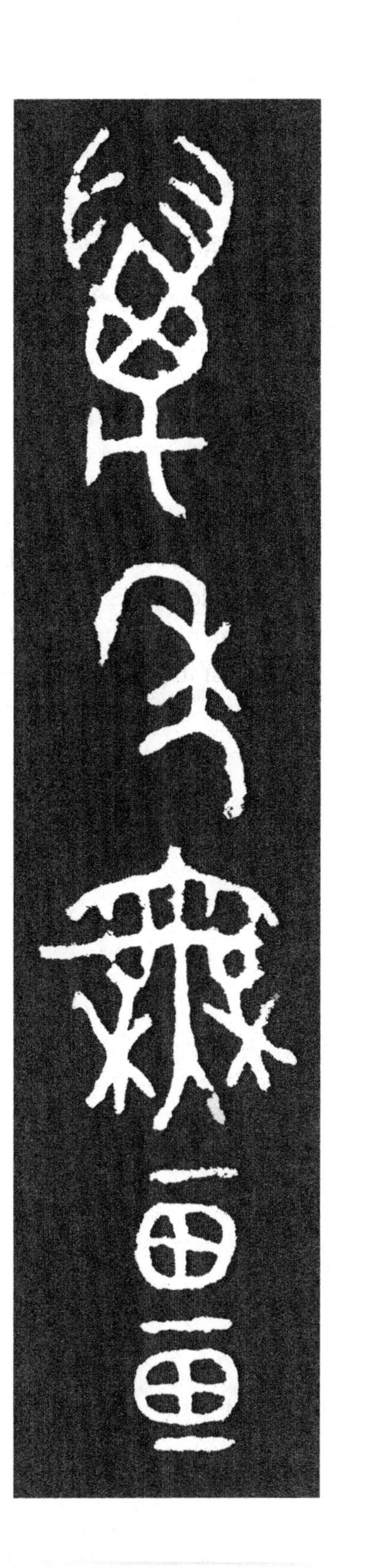

The general impression from the *Seventh Month* is that the life of the heavily exploited peasant, though harsh and circumscribed, was perhaps not altogether insufferable. And there were no doubt bright moments also, especially for the young, as seen in the next selection *Rouse not the Dog,* but this was from the more developed region further east in present-day Henan province and probably also later in time.

Rouse not the Dog

From Shijing

Upon the open field there lies a fallen hart.
To rope it up with straw is next the hunter's part.
That lucky fellow though is spending half his time
In leading on a girl with springtime in her heart.

With branches stout cut freshly from the forest's shade
Truss up, one ought, the deer there lying in the glade.
But come! What is a man to do with girl so fair,
With skin as gleaming white and smooth as polished jade?

"Calm down and take things slowly, do!"
"Do not mess up my apron so!"
"Rouse not the dog, you silly, you!"

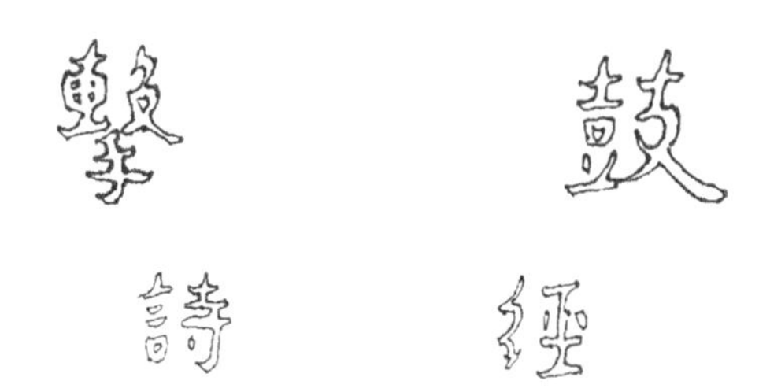

Politically, in the Spring and Autumn period, the Zhou kingdom was divided into numerous independent feudal states owing but titular allegiance to the king, who was himself in fact but one of the weaker feudal lords. Like feudal lords everywhere, these were continually engaged in war against one another to extend their influence, but it seems that their martial activities were mostly limited in scale and conducted along chivalrous lines. The aristocracy fought in horse-drawn chariots with bows and arrows and with lances. The common foot-soldiers, conscripted from the peasantry and poorly armed, had seemingly but a minor role. Indeed, the strength of armies was then gauged not by the number of men but by the number of chariots each put into the field. In principle, the peasants were enrolled only after the harvest so as not to disturb production and, after the event, were returned to the land. Some less fortunate, like the author of *War-drum* cited here, perhaps for some unusual circumstances as indicated by his possession or charge of a horse, were drafted for longer periods. The events mentioned probably refer to some campaigns undertaken by the Marquis of Wei in 597 B.C. against his neighbours in present-day Henan. (See map, Appendix I.)

There was, however, a more serious dimension to the military activities of the Zhou feudal states which was not often highlighted in the chronicles of that period. The nations in the north and west, with which the Middle Kingdom then had most contact, belonged to an ethnic group called by Chinese chroniclers the *Dili,* or *Di* for short. According to modern research, this was just an inaccurate tranliteration in non-phonetic Chinese of the name *Türk,* meaning strong, by which this people called themselves, which people, with appearance similar to the Han Chinese but speaking a different language of the Altai family, was later to spread widely over all of north-central Asia.

War-drum

From Shijing

Bang, bang, bang, the war-drum sounds.
We practise arms with leaps and bounds.
Just I got sent off to the south.
Others but build on near home-grounds.

In command of Sun Zizhong,
We first have conquered Chen, then Song,
But still they would not send me home,
And though disgruntled, must I along.

O when at last they'll set me free?
Or when perhaps unhorsed I'd be?
Then where my love would find my bones?
Beside some wood? Beneath a tree?

"Together, we, through death and life,
Nevermore to part as man and wife,"
We vowed then as your hands I held,
"Both growing old in peace, not strife."

Alas, they've kept us far apart.
I'm given no chance to play my part.
The gulf between us 's wide and deep.
I'm given no chance my word to keep.

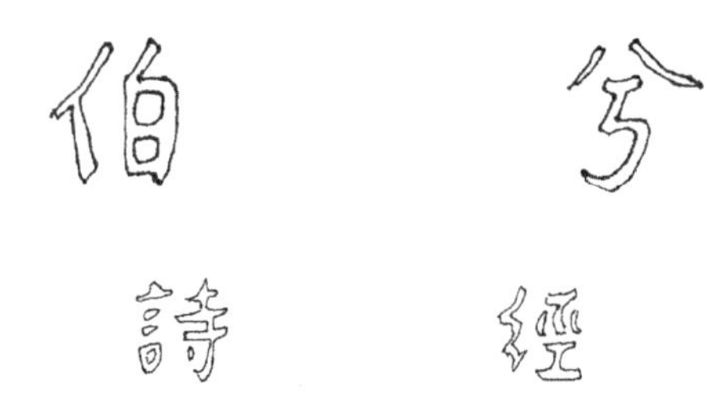

The two peoples were ancient neighbours, but as the Middle Kingdom advanced in culture under the Shang and Zhou dynasties, it became aggressive, as advanced cultures usually do, and pushed the *Di* to the less hospitable regions further to the north and west. In mid-seventh century B.C., however, those *Di* nations remaining nearby were united temporarily under a strong leader, which put the Middle Kingdom under serious threat, so much so indeed, that at one stage, according to a near-contemporary chronicler, its continued existence was hanging merely by a thread. It took the Zhou feudal states strenuous efforts, which were sometimes even concerted, to save at length the situation. For this reason, Confucius (Kongzi) who came about a century afterwards said of Guan Zhong, a minister of Qi who masterminded one of these efforts, "Without him, I would now be wearing my hair loose, dressed in a tunic fastened on the left", that is, do as those barbarians do. The threat eventually subsided but the *Di* elements continued to play a significant role in history well into the Warring States period, then dissolved eventually into the great melting pot of Middle Kingdom culture and became its citizens indistinguisahable from any others

The military exploits of the feudal lords, whether in defence of the Middle Kingdom or against one another, were, of course, amply recorded in history, but so long as they did not affect one directly as they did the poet of *War-drum,* or the poetess of the next poem *Fluff-ball* who was left behind, they were of little interest to the common composers of folk-lyrics, whose main concern, no doubt, was seeking the common necessities of life, as in *Seventh Month,* and if they were young, love.

Fluff-ball

From Shijing

O my love, he is so big and strong,
That he stood a head above the throng.
And the way he boldly held his pike,
As they marched before the King along!

But to the East since now my love has gone,
My poor, poor hair is like a fluff-ball grown.
Lotions I have, of course, and unguents sweet,
But what's the sense, my trimming up, alone?

We have been waiting, waiting for the rain,
But we have got the scorching sun instead.
And I've been longing, longing for my love,
To end up just with aching heart and head.

Day lilies help one to forget, they say.
So I'll find some and plant them by the wall.
For all this longing, longing for my love
Has turned my once-sweet heart to bitter gall.

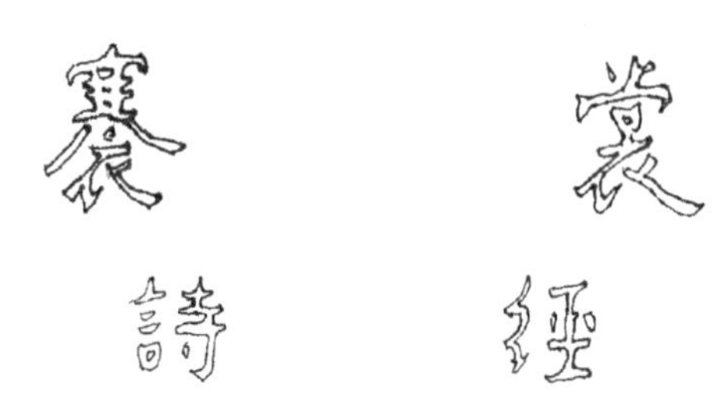

Apparently, young people were allowed to mix and flirt quite freely as seen in the next poem, much more so than in latter-day China when social structures became more rigid. *Tuck up Your Clothes*, however, was from Zheng which was then known for frivolity. Situated in present day Henan right in the midst of the then known world where trade routes crossed, Zheng was prosperous and cultured, with the aristocracy tending towards decadence. But the image of our poetess here is still one of perfect innocence. Even Confucius, reputedly one of the editors of Shijing and always mindful of proper behaviour, could say of the whole collection: "Three hundred poems described in one simple phrase is 'Think no Evil'." We would probably agree.

Indeed, by the time of Confucius in the sixth century B.C., Shijing had already taken more or less its present form and was appreciated and held in great respect. So much so, in fact, that an occasional quotation from Shijing, even if vastly out of context, was considered a necessary accoutrement for expressing oneself in polite society. For instance, the line *Rouse not the Dog* from a previous poem was once quoted in a diplomatic exchange, to warn an erstwhile friendly neighbour against aggression. Indeed, it is perhaps partly due to this high respect held then for Shijing that the anthology has survived to us largely intact.

Tuck up Your Clothes

From Shijing

If you'll be so good as to think of me,
Then tuck up your clothes 'n wade across the River Zhen.
But if you do not have me always in your thoughts,
Why then — will there not be another one?
You silly, silly silly silly boy!

If you'll be so good as to think of me,
Then tuck up your clothes 'n wade across the River Wei.
But if you do not have me always in your thoughts,
Why then — will another not look for me?
You silly, silly silly silly boy!

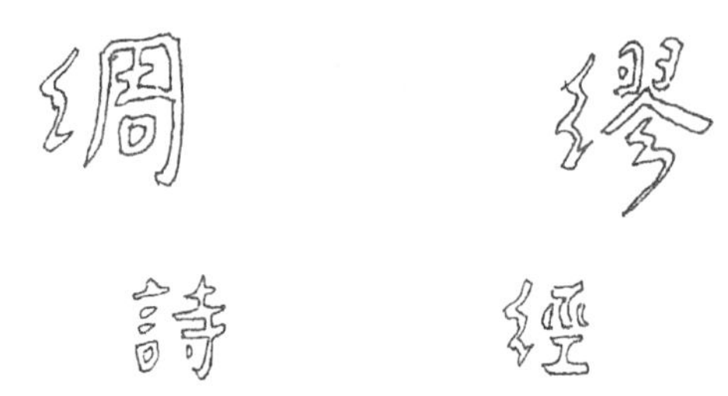

Technically, Shijing as poetry was perhaps a little primitive compared with what was later achieved. As explained in the Introduction, the Chinese language has tones, which can be employed by the poet in addtion to rhythm to add force and nuance to poetic expression. A Chinese poem is thus not just to be read or recited but to be half-sung like a song. Indeed, in Shijing's days, poetry and music seemed not even to have each a separate existence. The poems in Shijing almost all have uniform short lines of only four characters each, which do not always allow the tonal quality to be fully developed. Verses are often repeated with just small changes as in the next poem *Tightly Bound,* to indicate the passage of time. The rhyming is sometimes erratic and the rhyming pattern often irregular. It is indeed a good measure of Shijing's greatness that such poetry had been achieved with only these primitive tools at its disposal. In the English rendering here, of course, the tonal quality of the original is lost, but apart from the *Seventh Month*, the same rhyming patterns have been maintained.

In *Tightly Bound*, the poet seemed to find in his bundles of firewood a good omen for the fulfilment of his coming union with his beloved.

Tightly Bound

From Shjing

In bundles tight shall I the firewood tie.
Orion bright is up there in the sky.
Tonight — Ah, what a night! — that I should meet
With one so fair, so pleasing to the eye.
O boy, my boy! and what are you to do
With one so fair and pleasing to the eye?

In bundles tight shall I the dried grass stove.
Orion's by a corner of the roof above.
Tonight — Ah, what a night! — that I should meet
With her who loves me, whom I dearly love.
O boy, my boy! and what are you to do
With her who loves you, whom you dearly love?

In bundles tight shall I the thorn-bush store.
Orion now appears beside the door.
Tonight — Ah, what a night! — that I should meet
With that most brilliant girl that I adore.
O boy, my boy! and what are you to do
With that most brilliant girl that you adore?

However, not all marriages ended happily, as witnessed by the heroine of *Winning Smiles* cited next.

All these things happened to individuals somewhere in China more than two thousand five hundred years ago, but were depicted so vividly by Shijing that they still resonate across time and space and engage our full sympathy.

The world around these people, however, was changing fast as the Spring and Autumn gave place to the Warring States period. The previously numerous petty states were gradually coalesced, by increasingly frequent and vehement wars, into only seven. And these wars were no longer the almost amateurish affairs fought among the aristocracy, with the common foot-soldiers playing but a supporting role, but wars fought between field armies in which the common soldiers bore the brunt of the losses, and from these no individual, whatever their social status, could any longer be insulated.

Winning Smiles

From Shijing

A youngster then with winning smiles
Who traded here in silk and cloth.
But it was not for silk you came
That time; it was to win my troth.
I went as far as Hill Dunqi
To see you cross the River Qi.
"It wasn't I put back the date;
Matchmakers, they, could not agree.
Do not you, love, then angry be.
By autumn we'll each other see."

High-perched on top that broken wall
I watched from far the customs-stall.
When you for long did not appear,
My anxious tears began to fall.
But when at last you did appear,
We laughed and talked, as I recall.
"Go, seers and augurs now consult.
If no bad omen should befall,
Then send your wagon soon along
To fetch me and my dowry, all!"

Before the mulberry shed its leaves,
It was a most luxuriant hue.
But birds, forbear its berries sweet,

For these, they say, will fuddle you.
So girls, forbear the love of men,
Or soon the very day you'll rue.
For these men if they fall in love
Are able soon to shake it off,
But we girls, should we fall in love
Are able never to shake it off.

And when the mulberry sheds its leaves,
They turn yellow before they fall.
In poverty, since I left my home,
Lived I with you three years in all.
So now across the Qi again,
Its waters wetting the wagon's train.
'Twas not the wife who lacked in faith,
So had the man in faith remain.
O, you men can never stay on track,
From keeping faith you soon turn your back!

For three long years as wife to you,
No easy task had I to do.
Early rising, late to rest,
In household chores I did my best.
When things at last were put to order,
Sunny airs had turned to thunder.
My brothers, they would only smile,
Not understand, nor want to bother.
In silence thinking now alone
I'll grieve, no sharing with another.

"Together we'll grow old," said he;
Only bitterness he's left for me.

Two banks it has, the River Qi;
So Love an end must also see.
Recall the feast I came of age?
How full of talk and laughter we?
And you of oaths and promises?
Those days again will never be.
But what is the good of looking back?
That was the end of that, alack!

The Warring States

The Warring States period, as the Spring and Autumn period before it, saw a great advance in, among other things, science and technology. As technology developed, production was improved to support an ever larger population, but unfortunately, as always, so were the means of destruction to kill and maim a sizeable fraction of it. High quality steel was used to make ever sharper and more durable cutting edges, and advanced engineering to make crossbows with a range of up to six hundred paces. Against such new weaponry, the heavily armed aristocracy lost much of its advantage, in much the same way as the armoured knight in Europe was said to have lost his centuries later at the battle of Crecy against the English invention of the longbow. The chariot became obsolete; cavalry was introduced. Warfare was modernized, with the common soldier doing most of the fighting, and war became vastly more destructive.

By the mid-fifth century B.C. when the Warring States period officially began, the seven states which remained were Qi, Chu, Qin and Yan on the periphery respectively in the east, south, west and north, plus the the three central states Zhao, Wei and Han. At the beginning, they were all more or less equal in power, with each capable of putting an army of several hundred thousand into the field at any one time in the continual struggle for predominance. If anything, it was the central states which held then the advantage, occupying as they did the region which was both culturally and technologically more advanced. However, being hemmed in by the others, they had little room for further development, and soon began to lose out. Of the others, Qi was bounded on the east by the ocean and Yan by the barren north, which left Qin in the west and Chu in the south, each with a vast hinterland with rich resources to exploit, gradually to emerge as the only serious contenders. Chu, however, was still run along feudal lines, while Qin had for some time already developed an effective central government. Besides, Qin was strategically placed in present-day Shaanxi, protected to the east by mountains with easily defensible passes, which allowed it to retreat to safety when under threat. (Indeed, it was this same strategic advantage that prompted, two and a half millennia later, the epic Long March of the Chinese Communists under Mao Zedong to Yan'an which led eventually to the foundation of the People's Republic.) It seemed thus almost inevitable, with hindsight, that Qin would end up as the winner.

The battles in the war for dominance were fierce and brutal, and the losses were often truly horrendous. In one engagement, for example, in an attack by Qin on Zhao, the latter kingdom was recorded to have lost four hundred thousand men, which bled it of adult males for a whole generation.

Despite these devastating wars, however, society, it seems, continued to progress. The area under Han ethnic rule expanded, both northwards to the borders of the Gebi (Gobi) Desert and southwards all over the great Changjiang (Yangtze) river basin, nearly trebling the territory occupied in the Spring and Autumn period. The economy boomed, trade flourished, and great cities grew up. The capital city Linzi of Qi, for example, where "men rubbed shoulders and carriages their hubs", was said in a contemporary record of having seventy thousand households, which reckoned then at three adult males each, meant a total population of at least half a million. Schools of thought also flourished and became more specialized. Besides the seekers for universal truths such as Kongzi (Confucius) and Laozi (Lao Tze) of the Spring and Autumn period, there were now specialists in, among other subjects, economics, military tactics, logic and law. Indeed, it was the adoption by Qin of the doctrine of the so-called Legalist School that a country should be governed by codes of law rather than by the whims or benevolence of the rulers that gave Qin the final winning advantage over its rivals. Poetry, on the whole, however, seemed to have suffered, at least as judged by what is left to us, which is far inferior to what was recorded in the Shijing of the Spring and Autumn period, apart, that is, from that one shining beacon, Qu Yuan, of exceptional brilliance.

Qu Yuan (343–278 B.C.) was the first great Chinese poet to have left behind a body of work bearing his name. Born near the end of the Warring States period, the poet saw his own native Chu first weakened and then invaded by Qin, its powerful neighbour, and his poetic work is seldom far from his preoccupation with this theme. Of noble lineage himself and recognized already in his youth for his intellectual gifts, he tried, while in favour at court, to turn the tide by initiating reforms, but incurred thereby the hostility of the nobles whose privileges these reforms threatened, and eventually through them also the displeasure of the King. His last years were spent in exile to the south, where in despair after the fall of Chu's capital to the invading Qin army, he threw himself into the river and drowned.

Qu Yuan's poetry draws on two rich deep veins, first, the colourful local traditions of his native Chu, and second, the ancient lores of China's mythical past. Thus, based on Chu folk lyrics, he created a new poetic form very different from all that had gone before. In place of the terse, almost cryptic lines of four characters each prevalent in Shijing, he favoured generally much longer lines punctuated in the middle by a pregnant pause marked by an interjection. The result is a haunting tonal quality which, though often copied, has never been surpassed. Later, in *Lament for the Fall of Ying,* I shall attempt to give an indication how this device works. Further, like other ancient civilizations, China presumably also had a long and varied mythical tradition, but for some reason, little of it survived. Qu Yuan, however, made much use of whatever still remained at his time. Indeed, some of what is known today survives only through his poetry, for example in *Tianwen* (Questions to Heaven) from which the following entry is extracted.

In the Beginning there was...

Qu Yuan

O who can tell us of the far, far distant Past when Time was young?
What meant the question: "Up or down?"
ere Earth did first from Heaven part?
When Light and Darkness lay yet mingled,
what distinguished each from each?
Since all around was formless still,
so from which point could knowledge start?

Then Light to follow Darkness, Darkness, Light —
what purpose did it serve?
And Yin to merge with Yang the world to form, —
how functioned, how began?
And who could gauge the height of high-domed Heaven
layer upon layer?
Who was the prime creator who conceived this vast magnificent plan? ...

His poetry recorded also the local Chu traditions which differed in many ways from those prevalent in the rest of the country. If these traditions ever existed in northern China, they had left by Qu Yuan's time little trace. They could thus be ancient relics from before the foundation of the Zhou dynasty seven centuries ago, or else they could be peculiar to the Chu region. In either case, they underline the diversity of origin of what we now know as the Chinese culture.

He Bo, *the God of the Yellow River,* cited next, belongs to a group of eleven poems known collectively, for some reason never entirely settled among experts, as the Nine Songs *Jiuge.* They appear to be adaptations of folk ceremonial hymns sung in the area watered by the Rivers Xiang and Yuan (in present-day Hunan province) to the local pantheon in Qu Yuan's days. Like the Greek gods, the Chu deities have passions like our own, and many of the poems deal with the relationship between gods and humans in which religious and amorous fervours are inextricably mixed. Presumably, being but primitive deities, so would be their passions and their tastes, but under Qu Yuan's handling, they become endowed with both human sensitivity and celestial grandeur. For instance, the custom of sacrificing every year a virgin as bride to the Yellow River is known to have survived long after Qu Yuan, and it is not agreed among experts whether the poem refers to this barbaric custom or to the more idyllic myth of a romance between He Bo and the beautiful goddess Fu Fei of Luo, a tributary of the Yellow River. In any case, for the god in the poem, only the divinity remains, not the barbarity. The tonal beauty of the poem is superb, which unfortunately the translator found impossible to reproduce.

The God of the Yellow River

Qu Yuan

"Come, let us wander o'er my ninefold River Realm,
To be greeted by the wind and surging waves.
(Then let us wash and dry your hair by Xianchi Pool,
Where the sun at setting comes and nightly bathes.")

Lotus decked, his chariot,
A dragon, his every steed.
(For his Love he's waited long
But still she pays no heed.
Against the roaring gale he bursts into a song.)

"On top of Kunlun I survey the scene below.
Up soars my heart as if it would the world o'erflow.
And though the sun be setting, I entranced stay,
Enthralled by shores that shimmer in the distant glow."

Palaces adorned with pearls -
Whose roofs are fish-scale thatched.
Archways built of purple shells -
To halls by dragons watched.
How does the Spirit who within these waters dwells?

"Astride a giant turtle white,
In chase of fishes many-hued,

Come ride with me, my Love, my Sprite,
Over billows by me subdued."

Then facing east, her arms on breast so meekly crossed,
Our gentle maid, from southern shore is led.
The towering waves are sent for greeting her,
And schools of fishes are to guide her -
To her bridal bed.

The lines in brackets above actually belong in the existing text of *Jiuge* to another poem in the collection, but some experts believe that they should by rights be here, being artistically out of place in the other poem but fitting the present one like a glove. The translator fully agrees with their assessment.

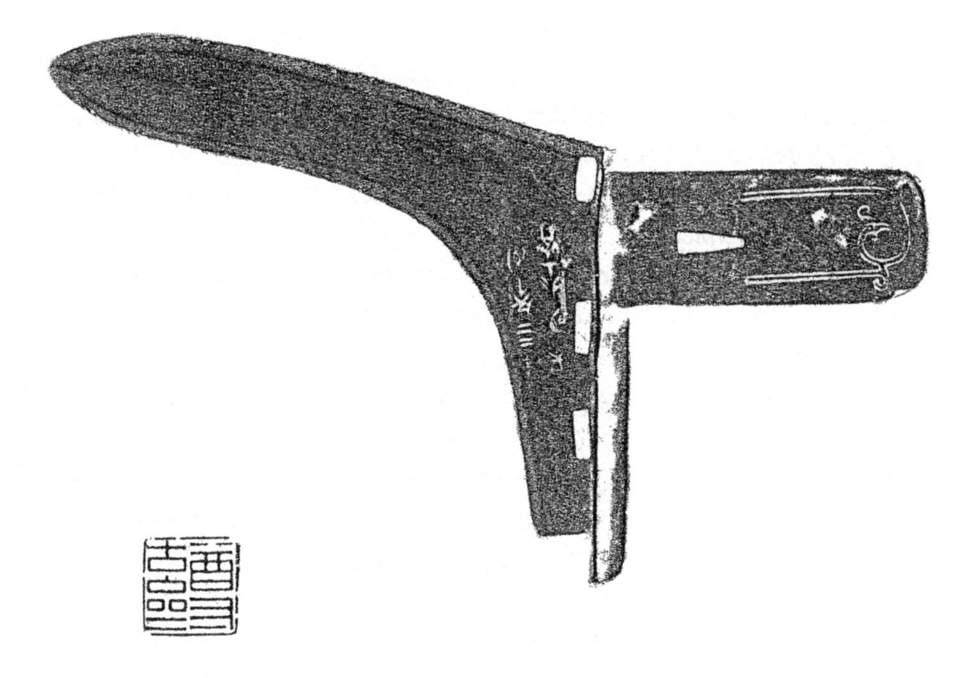

In spite of the truly haunting beauty of the *Jiuge,* however, it was with what one might call his patriotic poems that Qu Yuan had made the deepest impression on the Chinese mind. These include the great *Lisao* which runs to more than 350 lines, an exceptional length for a Chinese poem, and which is considered by many his masterpiece.

Though long exiled for no other reason than his efforts to save Chu from the disasters he saw coming, he had never till his death once wavered in his loyalty to the nation or, in terms of the perception at that time, to its sovereign. And this was in an age when it was the custom for men of ability to tour the country and offer their services for sale to any buyer, attaining thereby both wealth and high position in other lands, often to the detriment of their own. His deep devotion to the nation, coupled with his poet's sensitivity and, of course, his consummate skill, produced in these poems a sublime quality, tragic certainly, but not without imbuing us too with exaltation in the indomitability of the human spirit against unsurmountable odds. For this reason, his eventual suicide appears to us not as a surrender to fate but as an ultimate sacrifice in a final attempt to arouse the nation to its own defence. Tradition has it that the custom still observed in most of China on the Fifth Day of the Fifth Month of preparing *zongzi* (rice wrapped up in bamboo leaves said originally to be thrown into the river as offering) is in memory of his death.

His *Aiying* quoted next, one of the most moving, was written soon after the fall of Chu's ancient capital Ying in 278 B.C. to the invading Qin army, which marked the beginning of the end for Chu as an independent nation.

The interjection rendered here as "ah" is pronounced in Chinese differently in different dialects, but is basically just an articulated sigh.

Lament for the Fall of Ying

Qu Yuan

High Heaven's mandate is, I know, conditional — ah,
But why, against the people too, such heavy hand?
Bewildered they, and scattered, lost to one another — ah,
At just mid-spring are driven from the native land?

So too must I our ancient capital relinquish — ah,
Along the Rivers Jiang and Xia have I to flee,
With breaking heart to pass once more the city's portals — ah,
On this Jia morning, ne'er again the same to see.*

To leave thus Ying, and all in it that I have cherished — ah,
Distraught in mind! Will this come ever to an end?
Though oars be raised, our boat it falters on its journey — ah,
In grief that on my liege no more these eyes will bend.

With heaving sighs, and tear-drops fast like dense hail falling — ah.
I watched our tall catalpa tree recede behind.
Then round the headland on to Xia, and westward floating — ah,
Searched I by eye the Dragon Gate, but not to find.

A failing heart as I begin my aimless wander — ah,
I hardly knew just now on what I've put my feet.

*The first day of a ten-day "week".

But had I known thus drifting with the current — ah,
Would I've been exiled, and they like dirt me treat?

So now, where shall I go when sudd'nly sent off flying — ah,
Thus o'er, and like, the waves, the flood's endless futile crop,
My heart a knot that's never more to be unfettered — ah,
Through which course tortuous thoughts that never stop?

But soon, our boat is turned and we are floating downwards — ah,
The Dongting Lake's above, beneath us th' River Jiang.
Then east, forsaking home that we and ours have lived in — ah,
Since early times when Chu began, for centuries long.

Alas, my soul, how it longs now to be back there — ah.
Will any moment ease this longing to return?
The more my yearning grows for Xia my loved homeland — ah,
The more towards the West my back away I turn.

A high sandbank I climb and gaze into the distance — ah,
To try to ease awhile my aching heart in vain,
And see in sadness that peace yet reigns this distant corner — ah,
Where down the River, Chu traditions still remain.

Whence comes this torrent sweeping everything before it — ah,
And me too to the far, far South — but where away?
And who could have imagined glorious Ying in ruins — ah,
Its two great Eastern Gates abandoned to decay?

The loss of hope to ever cross again this River — ah,
Adds nation's future fears to mourning for the past.
As endless is the road that leads me back to homeland — ah,
Till all enternity will then my sorrow last.

And to recall that by my sovereign I've been banished — ah,
For many years I've been forbidden to return.
His sudden loss of trust in me for no good reason — ah,
Has rankled in my heart, where doubts forever churn.

They seemed so brilliant, to every wish they pander — ah,
But had at depth no strength on which one can depend.
Devoted men like me they barred from royal service — ah,
For to the jealous, both loyalty and faith offend.

Thus even Yao and Shun of old, the sainted emperors — ah
Though faultless their conduct, Heaven-high their fame,
Have yet to suffer vicious tongues who would, through envy, — ah,
Accuse them of unkindness and soil their burnished name.

Integrity and faith of good and honest men are hated — ah,
The blandishments of loose and empty men admired.
A jostling rabble fills the court, the good are banished — ah,
Is it such wonder that what has come to pass transpired?

And so,
With dazed unseeing eyes I look around me — ah
Still hoping once that I'll be back — but when?
A bird will fly great distance for its homeland — ah
A dying fox still heads towards its den.
'Twas truly not my fault to be so exiled — ah,
By day or night could I forget it then?

The Qin Dynasty

The Warring States period ended in 221 B.C. with China united under Shihuangdi, the First Emperor of the Qin dynasty, known now the world over by the terracotta army which guarded his tomb. The Zhou rulers styled themselves only as kings *(wang),* which title was in any case already usurped by all the rulers of the Waring States, and was clearly no longer grand enough now for the lord of the whole known world, who had thus to assume the title of emperor (*di)* originally meant for the King of Heaven.

Ruled from its capital Xianyang near the modern city of Xi'an, the empire extended from about the eastern boundary of present-day Xinjiang and Tibet eastwards to the sea, and from the Gebi (Gobi) Desert southwards into what is now the north of Vietnam. Although the empire lasted but fifteen years, it left an indelible stamp on China.

Discarding the feudal system of Zhou, Shihuangdi divided the country into provinces and counties governed under the same code of law by centrally appointed officials answerable for all their actions to the emperor. Weights and measures were standardized. A uniform writing system was introduced replacing the many scripts in use during the Warring States period, which, though having all a common origin in the oracle bones of the Shang dynasty or earlier, had developed wide regional variations. This last achievement was possible because the Chinese script is pictographic rather than phonetic, so that the written word can be understood across wide separations in both time and space, counteracting the diversification in pronunciation and usage inevitable over an area as large as that of Europe and a history at least as long. Such measures all helped to unify China, which is from then on but one nation. History would yet often see China divided into smaller political units in the centuries which follow, but this would be regarded as an abnormal situation, only rectified when the country is reunified.

Shihuangdi intended his empire for eternity, and to ensure that it should belong to him and his forever, he ruled it with an iron hand. To defend the empire against nomadic invaders from the north, the Great Wall was built joining the earlier structures of individual Warring States to form a continuous bulwark. And to safeguard the inheritance against subversion from within, weapons and even tools of metal were gathered and impounded in the capital to deny potential rebels the means of gaining or forging arms. Dissenting scholars were slaughtered and their books burnt in order that no idea inconsistent with Qin rule could ever be propagated. (At least, tradition had it so, although modern scholars tend to doubt the authenticity of this last accusation.)

Yet only shortly after his death the empire already started to crumble. The rebellion began with just a small band of conscripts who had missed their schedule because of rain. Since, under Qin's harsh laws, missing a schedule for whatever reason was punishable by death, they had nothing to lose. The insurrection started with only "bamboo stems for banners and sharpened stakes for weapons". Qin rule, however, had by then become so unpopular, that this tiny spark was enough to light a conflagration engulfing the whole country and in a few more years brought down the empire.

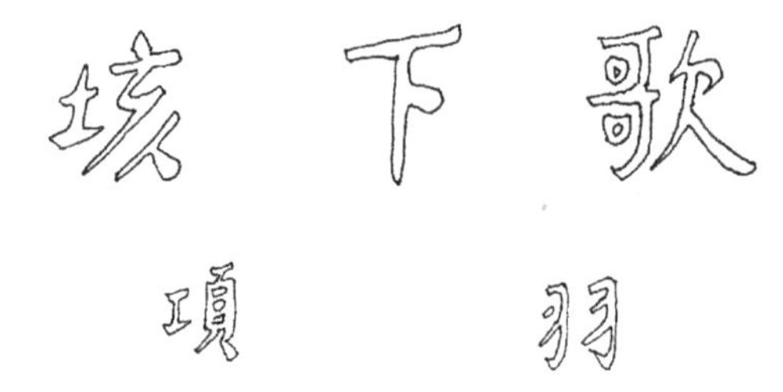

Xiang Yu (c.232–202 B.C.) was but in his early twenties and had neither name nor fortune when the rebellion began. In just three short years, however, he had by sheer force of personality and prowess made himself the head of the insurgent army and led it successfully to topple the Qin empire. He then became the new power afterwards, dividing the country as he saw fit into kingdoms for his lieutenants, with himself at the head as the Chief of Kings. His rule, however, lasted only five years. Brutal, with scant regard for human life, he inspired fear in all his opponents. Yet, to those close to him, he was strangely warm-hearted, even sentimentally so, and gained deservedly their unflinching loyalty. Such qualities won him the empire but were not enough to secure it. Thus, though brilliant at war — having fought, as he boasted, more than seventy battles and lost not a single one — he was useless in statecraft, and proved no match in manoeuvring to his wily arch-rival Liu Bang who managed in the end to turn the whole country against him. Finally, he was encircled on all sides at *Jiulishan.* On his last morning, unsure of what he should do, he reputedly composed and sang the following poem for his mistress, Yu. Yu, however, had her own ready answer to the question, and after dancing for him one last time, killed herself to set him free. He did manage later, against all odds, to break out again with a small following, but could not go far before he was again surrounded by the banks of the *Wujiang* (Black River). There, after a further spirited defence, and still unconquered, he took his own life. His last stand has since passed into legend and become an inspiration for drama.

Like Qu Yuan, Xiang Yu was from Chu, hence the occurence also, though not so subtly, of the mid-sentence interjections in his poem.

Encircled

Xiang Yu

Strength to uproot mountains — ah,
Zest to daunt the world entire!
With Fortune now against me — ah,
Even my steed has lost its fire.
When even my steed has lost its ardour — ah,
What is there that I could do?
And what, O Yu — ah — O Yu, my Love —
And what am I to do with you?

The Han Dynasty

In contrast to Xiang Yu, with neither his charisma nor his warmth, but scheming and exploitative, Liu Bang seemed by near-contemporary descriptions to be unattractive at the personal level. He was, however, more humane (or perhaps just less brutal) and had a statesman's vision that Xiang Yu never had. For this reason, he not only succeeded in wresting the empire from Xiang Yu but went on to found the Han dynasty which was to last for over four hundred years.

Apart from being one of the longest in duration in Chinese history, the Han dynasty was also in many senses one of the most successful. Internally, the country was tolerably well governed, with in particular the two consecutive reigns of Emperors Wendi and Jingdi (179 – 139 B.C.), when tax on farmers was reduced to only one-thirtieth of their produce, being long remembered as examples of wisdom and benignity. In practical terms, China under the Han dynasty was peaceful and productive enough to have supported a population unprecedented in size. A census taken in the reign of Emperor Pindi around the beginning of the Christian era when the dynasty had already declined from its apogee gave a total of 13 million tax-paying households. Reckoned usually at four to five persons per household, this gives a population of around 60 million, which is comparable to that of the Roman Empire at its peak about a century later, or to that of Britain or France today. It was a figure not to be equalled in China, let alone surpassed, until nearly six centuries later in the Tang dynasty. Externally, Chinese influence was extended by trade and force of arms, the latter especially under Emperor Wudi. In some directions Chinese power was pushed beyond the boundary of modern China, to the east into present-day Korea, to the south Vietnam, and to the west Kazakstan, Kyrgyzstan and Tajikistan. Envoys were sent in all directions, reaching India, Persia, and Roman Alexandria in Egypt. Of course, this influence later shrank when the dynasty was in decline.

Perhaps the most durable achievement of the Han dynasty was its establishment of China as a nation, so much so indeed that the Chinese language and the majority ethnic group which speaks it both still bear the dynasty's name. It was Qin of course which first fused China into a single political unit with a common written script and a unified system of weights and measures, but this was at first but a conglomerate of the old feudal states held together by force. Only under Han did the country first genuinely feel itself a single nation.

Perhaps it was the popular nature of the uprising which overthrew Qin that first brought about this change. Previous to that, the rulers of the feudal states, from one of which the Qin empire itself developed, belonged almost exclusively to the aristocracy descended along ancient lines, or at least, so it was claimed. The people's station in life was only to obey. The revolution which deposed Qin, however, was ignited and led by ordinary people. Chen Sheng, who started it all, was merely a conscript from among the dregs of society, and many who joined in later were no better. Even Liu Bang who later became the August Ancestor *Gaozu* of the Han dynasty was but a minor county official when he too took the plunge. Thus, although the empire was still conceived to be the personal property of the emperor, his right to rule was no longer just mandated by Heaven, for clearly the people had had some say in it, though the means be violent and the purpose unclear.

With unification came the recognition of the whole country as a unit sharing a common culture, both material and spiritual, which was different from, and in fact at that time vastly superior to, that of its neighbours from whose encroachments the country had to be defended. And so the concept of a Chinese nation was born, not just among thinkers as previously but now among ordinary people as well.

In taste, as reflected in poetry, the Han dynasty was simple and pragmatic, with neither the exuberance found in Shijing before it nor the brilliance later achieved in Tang. Yet, it is in exactly these qualities that lies its strength.

The sorrow at parting and the longing for home are a constant theme for the poetry of the period, as in the *Cartwheels* which follows. No longer bound to the land as in former days, the common man could now travel in principle over the whole length and breadth of the empire to engage in business and trade. More often, a man could be drafted into the army and sent off to the frontiers. In either case, the outcome being uncertain and the distances great, the traveller might be away for many years and might never even return.

Cartwheels

Anonymous

Instead of shedding tears, I'll sing a mournful ballad,
Of home-return, I can but towards the distance gaze.
Yet still for long lost home my heart is ever yearning;
Bewildered and oppressed, my mind is in a daze.

But there's no boat to ferry me across the River;
No one awaits me at the home for which I yearn.
The thoughts within me, they can find no utterance,
But like cartwheels, forever and forever turn.

For the soldier departing to the frontier, the outlook is particularly grim. To the nomadic hordes roaming the vast reaches of northern and central Asia, the prosperous Han empire represented a glittering prize, if too strong to conquer outright, at least on occasion to raid. And thus to defend itself against them was one of the empire's primary preoccupations. It was to guard against these invaders that the Great Wall was built, and to garrison, strengthen, extend or keep in repairs these extensive ramparts required a constant draft of men from throughout the country. Conscripts were needed also for the distant and perilous expeditions into enemy territory which had to be mounted periodically to satisfy either genuine strategic needs or the martial ambitions of the reigning emperor. A soldier was drafted often for an unspecified period, and in some cases even for life, while conditions on the frontiers were hard, and skirmishes almost continual. It is thus not hard to appreciate the pathos at parting both of the departing soldier and of those he left behind, as depicted in *Parting at Dawn* below.

Parting at Dawn

Anonymous

I married you when both were young,
Between us e'er love and trust.
But 'tis the last of such our nights together,
The last of happy tender love.
A traveller's mind is on the journey.
I rose and peered into the night.
The stars of dawn have now all vanished,
Then it's away, we part forever.
From hence on battlefields I'll labour.
I know not if I shall see you more.
Your hands in mine, and deeply sighing -
Your eyes are wet - we part for life.
Be brave and careful of your tender youth.
Remember, love, the days when we were happy.
If I but live I will return to you,
And if I die, I'll keep you in my heart.

The new found concept of nationhood was no doubt a source of pride to the Hans, especially when China was then much more advanced in development than any of its neighbours, whom it regarded as mere barbarians. To defend that nationhood, however, required from everyone a sacrifice of some personal liberty in the form of loyalty to the nation, or in the terms as understood in those days, of loyalty to the emperor, who was the executive head of the nation as well as its symbol. Against the emperor's wishes, however whimsical and unwise, a subject could argue and remonstrate, but the emperor could not in principle be disobeyed. Such a system ensured a certain amount of stability, as witnessed by comparison of China then with other nations, but in the long run could obviously not work, especially since the emperor in latter days, for various reasons, was often enthroned when a minor, and bred in the palace with no experience at all of the outside world. A soldier posted to the frontier had thus no easy task, having to contend not just with the enemy in front but also the lack of resources and proliferation of slanders behind his back. Even when the empire was relatively prosperous and revenue forthcoming, this was often squandered by an extravagant court or on over-ambitious campaigns. And examples are many of generals having succeeded against the enemy only to be recalled, imprisoned, or even executed for failing to bribe or pay respects to some favourite of the emperor who happened then to have his ear. And for soldiers of a lower rank who did the actual fighting, they had further to contend with incompetent commanders who had no better qualification sometimes than being, for example, a close relative of the empress dowager acting as regent. Thus, through no lack of prowess or courage on their part, they became food for *Vultures*.

Vultures

Anonymous

Fighting south outside the city,
Or fallen by the northern gate,
To die unburied and devoured
By vultures is the soldier's Fate.

O please my friend, to the vultures say:
"Yet wail awhile for him — a little patience, stay!
The dead of war — he has no hope for burial;
Can carrion flesh then fly from you away?"

Deep and silent flows the river,
Its waters darkened by the floating weed,
While to and fro a packhorse whimpers
Around the carcass of a fallen steed.

When even the bridge is fortified,
How pass from one to the other side?
How gather then the millet and the rice
To feed you, m'lord, and by our liege abide?

O spare a thought for the loyal liegeman,
For well deserving of a thought is he!
The morning sees him off to battle,
The evening his return may never see.

Those fortunate enough to survive the battles were often kept on for garrison duty or to develop and farm the frontier regions, and would be lucky ever to get home again. Nor were those at home necessarily better off. In the latter days of the empire when the country had been greatly weakened and impoverished by reign after reign of mismanagment, any natural disaster would bring on famine and popular uprisings to be followed by imperial suppression often with even worse effects. *Home-coming* for the soldier was thus not always a happy event.

Home-coming

Anonymous

A youth of fifteen years I joined the army,
Near eighty old before I could return.

I met a stranger coming from our village.
"And who is left there living still at home?"
"Your home, my friend, is yonder in the distance."
A clump of stone-pines — silent as a tomb.

A pheasant flew from up among the rafters,
And rabbits hid where dogs once had a lair.
Some rice grew wild, the middle of the courtyard,
And greens on top the well in disrepair.

I'll glean and husk some rice to make a gruel,
And with some wild greens too, a broth I'll brew.
Both broth and gruel will soon be ready,
But who is there to share the meal with? — Who?

(As just a youngster I had joined the army.
'Twas many years before I could return.)
Hot tears stream down my eyes and drench my tunic,
When outside, towards the dawning east I turn.

The two lines in brackets have been added as a refrain by the translator just for taste.

What is impressive about the Han dynasty, however, which has left an imprint on the Chinese nation ever since, is its people's indestructible integrity despite all the misfortunes and abuses they had to suffer. *Even Hungry, I'll not Feed with Tigers* encapsulates in a few lines this spirit that has kept Chinese culture intact through several thousand years.

Even Hungry, I'll not Feed with Tigers

Anonymous

Even hungry, I'll not feed with tigers,
Though late, will not with fowls abide.
Has not the wildfowl each its chosen nest?
Then, homeless one, why this inordinate pride?

Even in their love poems, the Hans would stress more the obligations rather than the pleasures, the pathos rather than the joy. But here again we are impressed by their persistent, almost agressive integrity, as that of the heroine of the next poem in declaring her fidelity.

The above poems painted a grim picture of life during the Han dynasty, and they are quite representative of the Han poems which have come down to us. It seems unlikely though that the people who had sung so light-heartedly in Shijing before would have lost now altogether their joie de vivre. The picture also does not accord with what we know today of the Chinese people who mostly manage to smile and laugh even through adversity. Besides, if life was really so hard all the time it seems unlikely that the Han dynasty could have survived for its full four hundred years, supporting the large population it had. What seems more probable is that the taste in that period (which is in fact not uncommon even among Chinese people today) dictated that poetry be expressed only in dark colours, and that anything boisterous or even cheerful would be considered unpoetic.

Promise

Anonymous

Heaven be my witness!
I shall be true to you
All this long life through.
Sooner, love, this stream to lose its water,
Or that mountain there its peak,
Snow to fall in summer,
In winter, thunder speak,
Heav'n and Earth to crush together,
Than I to you my promise break.

Indeed, it is not often one finds in the existing literature examples of happy Han poems. *A Length of Silk* cited below belongs to a group known simply as *Nineteen Old Poems* of unknown authorship, and though still set in a background of long separation is already by far the happiest of the lot.

Technically, verse forms seemed to have evolved from Shijing's days and become more varied. The examples of Han poetry cited above are of two types: lyric poetry (Yuefu) sung with music, and poetry as an independent art form. The lyric poem, of which *Cartwheels, Vultures, Home-coming, Even Hungry – I'll not Feed with Tigers*, and *Promise* are examples, has generally lines of different lengths with three to seven characters in each, and an irregular rhyming pattern. A poem of the other type, on the other hand, such as *Parting at Dawn* and *A Length of Silk*, is more regular, with five characters in each line and rhymes every other line. In either case, the new forms allowed both tonal quality and rhythm to be more fully developed compared with that in Shijing days, and gave to Han poetry an impressive simple gravity which reflected distinctively the spirit of the time.

Most of the earlier Han poems which have survived were from unknown authors, which suggests that they were of humble origin, and their poetry had still the direct appeal that Shijing possessed. By the late Han dynasty, however, poetry became more and more a pursuit of the educated and, as a result, increased in sophistication and scope while losing in spontaneity. No doubt, the peasant still sang as he laboured in the field or flirted with the girl next door, but his songs were apparently not as appreciated as they had been before and often have not been preserved.

A Length of Silk

Anonymous

A stranger came from distant lands
And brought a length of silk for me.
My love, he has me still at heart
Though distanced by ten thousand li.
I think I shall make with it a duvet;
By mandarin ducks adorned it'll be -
And filled with silk like lingering thoughts[*]*-*
With knots be edged which'll ne'er work free.
Like glue, love, into lacquer mixed,
By nothing shall be parted, we.

* The Chinese character for silk (thread) has the same pronunciation (*si*) as the character for (tender) thoughts, although the two characters come from completely different roots. And the two also being both soft and seemingly endless, this coincidence is often exploited in poetry, as it is in the present poem.

The four hundred years' duration of the Han dynasty (206 B.C.–220 A.D.) was divided into two almost equal halves, called the Western and Eastern Han, with capital at Chang'an and Luoyang in present-day Shaanxi and Henan respectively. In fact, the dynasty was already overthrown by an usurper around the beginning of the Christian era and would have come to an end then, had not Liu Xiu, a scion of the dynasty, risen from the ranks, banking on the prestige of Han's earlier achievements, to give it a second lease of life. The fortunes of the two halves followed roughly the same pattern with a few good emperors at the beginning succeeded by the mediocre, then descending to the incompetent, the vicious and the ridiculous. Indeed, such a pattern was probably inevitable given the system in which the emperor held absolute power, whose ascent to the throne, however, except of course at the beginning, was based not on merit but on birth, and it was a pattern that would be repeated again and again in the later dynasties.

By the middle of the second century A.D., the Han imperial line had already reached the limit of degradation. The weak emperors with hardly a will of their own had effectively ceded the reins of power to their palace eunuchs, who had only their personal interests at heart. Driven to extremity, the heavily exploited peasantry rose in revolt, which when barely suppressed at one corner of the empire would break out again somewhere else. And as internal dissension grew and the strength of the empire declined, the raids of nomadic invaders from the north and west increased in frequency and virulence. So the Great Wall on which the empire's defence relied required more and more resources, both material and human, for its extension and upkeep. The following description of how the Wall was built and maintained, though given only by a poet **Chen Lin** (? –217 A.D.) of the upper class, seems nevertheless quite authentic.

The Building of the Great Wall

Chen Lin

Beneath the Wall for watering horses is a little lake.
Its water is so cold it seems the horses' bones would break.
I'll straight to Foreman go and boldly ask:
"Isn't it time to let us Taiyuans home, for goodness sake?"

"Official schedules, sir, are not your worry.
Your job is just to build while singing out in harmony!"

Ye gods! A man would rather die in fray,
Than labour at the Wall day after day, -
This wretched Wall that stretches on and on
A thousand miles, on thousands more, away.
It's kept out here the young and able-bodied,
While back at home but wives like widows stay.

I'll to my wife at home a letter write:
"My loved one, while still young remarried be.
Be good to your future parents-in-law,
But yet, for old times sake, spare occasional thoughts for me."

"Why such despondent thoughts, dear, so unlike
Your usual self, unworthy too of thee?"

"In penury, I, from which might ne'er return.
It's only right that I should set you free.

Raise not Baby, if a boy befell,
But if a girl, then raise and feed her well.
A boy, you see, will just be sent to build the Wall and suffer,
Where bones of dead and dying lie all jumbled — a living hell!"

"Since youth when first to you by marriage bound,
About you my every lingering thought is wound.
Full well I know that frontier life is hard.
You gone, you think I'll long alive be found?"

Even the Great Wall, however, could not keep out the invaders forever, and with them came pillage and devastation. **Cai Yan** (c. 172 A.D. - ?), daughter of Cai Yong, a much admired scholar and literary talent of the time, was captured by the invaders in one of these raids and carried away as a prize. Among the nomads up north, she was given to one of their chieftains and bore him two sons. Later, during a truce between Han and the invaders, she was ransomed by friends at a high price, brought back south and remarried, but she had to leave her sons behind. She stayed in the north twelve years. Being talented both as a poet and a musician, she was reputed to have composed a cycle of eighteen songs in the style of the hujia, a wind instrument of the nomads, adapting apparently the music to the Han qin, a string instrument. The following is a selection of these songs of which the lyrics telling her own story survive. Although scholars are not agreed whether the existing version is indeed the original one ascribed to her, it still gives a vivid picture of what she had to go through. Even then, given her elevated social status, she should not be numbered among the least fortunate in those unsettled times. These poems were written in a style rather like that of Qu Yuan, with long uneven lines each punctuated by a pause, very different from the standard Han format with uniformly five characters in each line. It seems that it is mainly for this reason that some experts doubt their authenticity. On the other hand, one can equally argue that in adapting to the nomadic music of the hujia, she found the standard Han format too restrictive and switched instead to the other.

Eighteen Pieces for the Hujia

Cai Yan

....

High Heaven, they say, has eyes - ah

Then why does it not see me driven off alone?

The gods, they say, are wise - ah,

Then why have they to far, far barren North me thrown?

I've Heaven never yet offended - ah,

Why has it banished me across the wastelands wide?

And nor have I the gods offended - ah,

Then why a foreigner given for my very own?

It was to ease my sadness - ah

That I this song, the Eighth, composed,

But now that it is finished - ah

The deeper sorrow in my heart has grown.

....

With glaring beacon-fires ne'er absent from the tower top,

The deadly strife on battlefields, when will it ever stop?

By day, the dusts of war choke passes barring me from home,

By night, above, a ghostly foreign moon the wild winds crop.

I forced back tears, though any news of home has long since ceased,

And tried my failing spirits for a while with music prop, --

Since life is plagued by parting - ah - what boots it to lament? --

But when this song, the Tenth, is over - ah - fresh tears by blood stained drop.

....

Life has for me no laughter - ah -
So Death should hear from me no groan.
There are but two things keep me - ah -
From what the fortunate bemoan.
So long as life is left me - ah -
There's still the hope one day of home,
But if once dead and buried - ah -
I'll ne'er for present pain atone.
Besides, my lord, though foreign - ah -
To me has some affection shown,
And from this forced union - ah -
I have two offspring of my own.
Without shame I them nurtured - ah -
And they are now to boyhood grown.
I cherish them, and pity - ah -
For they no better home have known.
Hence comes this Eleventh piece - ah -
Its mournful, soft but lingering tone,
Sinks deep into my heart - ah -
And chills the marrow of my bone.

....

A burst of spring brought warm fine weather even to this shore, -
The Emperor's beneficence, as in happy days of yore.
The nomad folk rejoice while joining all in song and dance,
For peace has dawned on both our nations; they will strive no more.
Then suddenly, meeting with the special envoy, I was told,
To buy my freedom there were sent a thousand pieces gold.
O joy, to go back home alive to face such blessed times!
But then, to leave my sons, and ne'er again their forms to hold?
This Twelfth, it is a mix of sorrow, joy in equal parts,
A whirl of bitter sweetness no word or music can unfold.

....

My thoughts were wandering - ah -
When I began this Sixteenth Song.
They're there now with my children - ah -.
Where they by rights belong.
Like sun and moon be parted - ah -
One east and one the west,
To gaze across the Endless - ah -
Each for the other long.
There's no means, no magic - ah -
Can help me to forget.
Like notes I'm fingering - ah -
My heart do pangs of sorrow throng.
It's true, I'm home, but - ah -
I had to leave my sons behind.
Exchange past suffering - ah -
But for a new one quite as strong.
With bitter tears down streaming - ah -
I raised my eyes on high.
Why give me Life, ye Heavens - ah -
If just to bear this wrong?

....

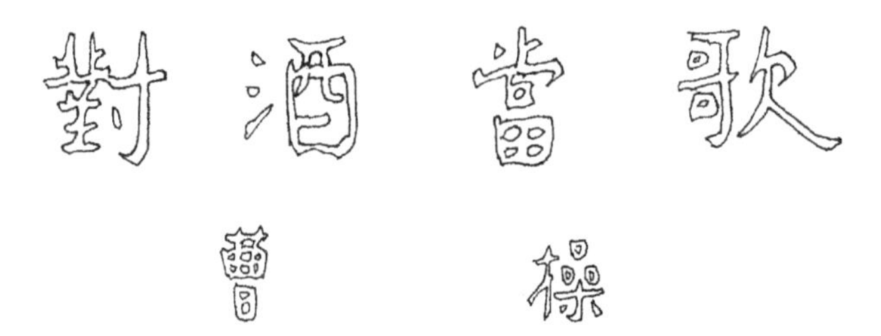

The Han dynasty ended in a spell of chaos, followed by the separation of China into the Three Kingdoms. Wei was by far the strongest of the three, holding sway over the then more populous and advanced half of China north of the Changjiang. The founder of Wei was **Cao Cao** (155 - 220 A.D.), a charismatic though ruthless figure - a brilliant general, astute politician and no mean poet. It was also he who paid the ransom to repatriate Cai Yan, the poetess of the last entry. Though having brought the major part of China under his control, he continued to pay lip service to the last Han emperor. It was only after his death that his son Cao Pi deposed Han and supplanted it by the Wei dynasty.

Despite his heavy schedule on campaigns and in government, trying to bring back some order to the empire in disarray, he still found time for poetry. Together with his two sons, Pi and Zhi, he started a new style of poetry based on that of Shijing and the Hans. Though keeping the formats, he imbued them with a new character and made of them a vehicle for personal expression. The poem *Wine! Let's Sing with it a Song* cited below gives a fascinating picture of his complex personality, combining the sensibility of a poet with the wiliness of a politician, and the limitless ambition of the adventurer with some sagacity of the true statesman. Though its exact date of composition is not known, it seems like a work at about the height of his power, when at the head of an army of close to a million, he was at the point of crossing the Changjiang southwards to complete his conquest of the country. With the unification of China in sight, he was inviting men of ability to rebuild the empire under his banner. The campaign, however, ended in disaster at the memorable Battle of Chibi, where Cao Cao was soundly defeated by a southern coalition army much inferior to his in number, and barely escaped with his life.

Wine! Let's Sing with it a Song

Cao Cao

Ah wine! Let's sing with it a song,
For human life, it lasts how long?
It's but a drop of dew at dawning.
Too much time has passed along.

And as for mem'ries and regrets,
The present cares one ne'er forgets,
What's there for their relief but brief
Oblivion that wine begets?

"My heart, it soft and tender grew
To see your tunic of dark-green hue,"
To wait till now with bated breath,
For whom, my friends, if not for you?

"Do not you, friends, the deer-calls heed?
They're out in fields on green grass feed.
And I have here my friends as guests!-
Give music, - Ho! - both string and reed."

O'er all the world her beams to send,
Ah when, oh when, will the bright moon end?
Can mortals, we, but envy her,
Whose lives but o'er an hour extend?

You came from far o'er hill and dale,
To grace my feast, myself to hail.
Recalling then our friendship past,
Let's at our ease ourselves regale.

The stars are scarce for th' moon is bright.
Some blackbirds on their southward flight -
Three times round that tree they flew,
But found no fit branch to alight.

It's, as the mountain's high as high,
Or as the sea is deep as deep,
By keeping faith with all the world
That Zhougong had its faith to keep.*

The lines in quotes are from Shijing, and Zhougong* was a statesman-minister of the Zhou dynasty traditionally revered as a role model.

Wei, Jin and the Northern and Southern Dynasties

After the final demise of the Han dynasty, there followed a period of some four hundred years during which China was seldom unified, that is, except for two short intervals, one near the beginning and one the end of the period, each lasting little more than a couple of decades. Weak dynasties followed one another, often in rapid succession, and although each claimed suzerainty, controlled but a portion of the country, sometimes no more than just a few provinces. The rest of the country was divided in a kaleidoscopic fashion into a multitude of autonomous states fighting with one another for dominance, or simply at times abandoned to chaos.

Most of the rulers, or rather warlords, in the northern half of the country were non-Han in ethnic origin. Some had been Han subjects, either by their own choice or through Han annexation, already for several generations. For example, some were descended from the Xiongnu nation, who once had a vast empire stretching from the Gebi (Gobi) Desert northwards into Siberia to as far as Lake Baikal, and rivalling the Han dynasty in power. When their empire split under Han pressure around the beginning of the Christian era, one half moved west reputedly to become, centuries later, the Huns in Europe, while another half surrendered and was incorporated into the Han empire. These newcomers to the Middle Kingdom and others like them were unfortunately treated rather shabbily, exploited for their military prowess and herding skill but never given full citizenship rights. So they bore a legitimate grudge against their Han rulers, and only waited for their chance to break free. Some others, on the other hand, were fresh invaders from the nomadic peoples of the north and west, who poured into the Middle Kingdom at first just for plunder, but when they found it only weakly defended, annexed for their own whatever they could. Being at first but half-civilized, some of them treated the local population barbarically, which barbarity, in revenge, the Hans sometimes returned when the occasion arose. Thus, coupled with the misgovernment in the previous decades under Han rule, the invasion took a heavy toll on the population which dropped at one stage to barely a fifth of that of the Han dynasty at its peak.

The original Han-ethnic ruling class, together with any who had the means to do so, fled across the Changjiang to the south of the country, which they managed to hold on to against the invaders. At one stage indeed, the fate of the south hung precariously in the balance when a formidable force from the temporarily united north said to be over a million in strength was poised for its destruction. But at the fateful Battle of Feishui (383 A.D.), partly by good generalship and partly by luck, the defending army only a tenth in size prevailed, allowing the south to survive and preserve its independence. There, dynasties of Han-ethnic rulers followed one another, and although few had the vigour even to attempt a return to the Middle Kingdom, they at least managed to maintain some semblance of the original Middle Kingdom civilization, which was bound in the end to attract the attention of the increasingly sophisticated rulers in the north from other ethnic groups. These began first to admire and then to emulate what the Hans had achieved in governance, and combining it with the vigour they had brought with them from their homelands, soon succeeded to build kingdoms better managed than those in the south. Consequently, it was from the north that the country was eventually reunified.

The period, though often regarded as the Dark Age for China when advances made in the previous millennia were suddenly and brutally arrested, was not without its positive contribution to shaping the China of today. First, it brought the non-Han ethnic peoples of the north and west into the central orbit and thereby eventually enriched the Chinese civilization. Secondly, it brought the know-how of the previously more advanced region north of the Changjiang to the south of the river, leading to rapid development of that fertile region, which was by then long overdue.

When Cao Cao died, the reins of government of northern China passed into the hands of his eldest surviving son Cao Pi, who did not have the same scruples as his father as regards the Han dynasty and soon had the last Han emperor abdicating in his favour. And so began the Wei dynasty.

The next poem by **Cao Zhi** (192–232 A.D.), brother to Pi by the same mother, is popularly known as the "Seven Steps Poem" for the following reason. When young, before Pi was named as heir to his father, Zhi and Pi were rivals, with Zhi generally considered as the abler of the two, and both were also accomplished poets. Thus, when Pi ascended the throne, he was both jealous of Zhi's reputation and wary of his ambitions, and sought, as the story goes, to have Zhi removed. One day therefore, he summoned Zhi to the imperial court to test Zhi's ability, in order to answer the allegation that some of Zhi's "brilliant" literary achievements had actually been pilfered from friends. If proved correct, the fault could be severely punished as perjury, since some of these compositions had been presented officially to their father, and some also to Pi himself as Emperor. After setting Zhi several tasks, yet failing to trap him in any, Pi finally ordered Zhi to compose, within the duration of just seven steps, a poem about their relationship as brothers, but without ever using the word "brother" in it. This poem was reputedly the result.

Though of doubtful authenticity, this anecdote illustrates well the fact that Cao Cao's descendants were not of his calibre. Indeed, given all the advantages left them, these not only failed to reunite China, but even found themselves often on the defensive against their weaker neighbours. Thus, after only a few generations, their inheritance was taken from them, in fact in much the same way as they had taken it from their predecessors.

Beans and the Beanstalk

Cao Zhi

Fry beans o'er beanstalk fire.
Beans weep inside the pot:
"From same roots grown be we -
Why scorch with flames so hot?"

Wei was replaced by the Jin Dynasty which finally succeeded to reunite the Three Kingdoms in 280 A.D., but only for a couple of decades. Weakened by centuries of misrule and internal strife, the Han nation could no longer withstand the invaders from the north and west who soon overran the northern half of the country forcing the Han ruling class to flee south across the Changjiang. For a while there was chaos, which after some years settled into a stalemate, with the region south of the River remaining under Han rule, but the north divided among rulers of mainly non-Han origin. None of the dynasties lasted long, however; in the south there were five in that period, and in the north even more, and every change of dynasty, of course, was an upheaval. Even by the beginning of the Jin dynasty before the full invasion began, a census showed that the population, that is, the countable part of it, had already dropped to barely a fifth of that of the Han dynasty at its peak, "comparable only to that of a large Han province".

In such a situation, it is not hard to imagine how the common people fared. Life would be particularly hard in the northern half of the country under the half-civilized war-lords each struggling by force for dominance and ruling their own portion with an iron fist. The people's suffering was reflected in the folk lyrics of the period, of which *Overnight Stop at Longtou* is an example. Longtou was in present-day Gansu province in the impoverished north-west. The style of this poem seems to have changed little from that of the Han period; indeed some scholars even believe that it may be a Han relic. But given the situation then it would not be surprising that development of poetic style was not a priority.

Overnight Stop at Longtou

Anonymous

The waters which from Longtou flow,
Left wandering on the mountain-side,
Remind me of myself, alone
To roam the open country wide.

From Xincheng, we made early start,
Pass night here at this spot remote.
It is so cold, I cannot speak,
With tongue sucked back into my throat.

The waters which from Longtou flow,
Why such a sobbing sound you make?
This distant view of Qin, my home,
Would be enough my heart to break.

The fall of northern Han China to the nomadic invaders, though devastating at the time, was not without its long-term compensation. Brought up in the harsher conditions back home, the invaders had come to central China intent only on plunder, but when they got there, they found that there was much to their taste in the advanced and settled Han culture. In *Dancing by the River,* for example, a non-Han fellow is depicted watching the Han people dancing; he longed to join in but could not manage the language. We can safely assume, though, that the time would come when he would have picked up enough phrases in the local dialect to chat up the girls. And though viewed with suspicion at first, he would soon be recognized as not such a bad fellow after all and be admitted to join the fun. On his part, he would be teaching them some of the dances and songs he had learned back home, including some brought there by travellers from the civilizations further west. We have then in the making a new and richer Chinese culture, for both Hans and non-Hans alike, which would come to bloom some centuries later in the Tang dynasty.

Dancing by the River

Anonymous

In the distance, by the Mengjin River bank, I see,
The Hans are dancing shaded by that spreading willow tree.
A pity though that I a non-Han fellow born and bred,
Just cannot understand what they are singing in their spree.

Meanwhile, of course, before such momentous developments could come to pass, some normal things of life would go on, like the interminable arguments between a teen-age daughter and her mother recorded in the next poem ...

Jujube

Anonymous

Outside our door there stands a jujube bold.*
It does not know it's yearly growing old.
Ah Mum! If you'll not have me married off,
You'll have no grandchild in your arms to hold.

* Tree with sweet date-like fruits and a tough useful wood.

... or a girl's preoccupation with her appearance, as depicted in the next.

The small number of poems from the Northern Dynasties which have come down to us are mostly folk lyrics of unknown authorship, indicative of their humble origin. They have thus still the same direct appeal of Shijing and of early Han. Presumably, those sophisticated poets of the higher educated social class who put their names on their compositions had already fled to the relative safety of the south and brought their art with them.

The Well on Top of Mount Huayin

Anonymous

The well on top of Mount Huayin's a hundred fathoms deep.
Its water is so cold, it makes the bones inside you creep.
A pretty girl can down the shaft at own reflection peep –
"Can't see the rest but just — but just can see my neckline steep."

Serving roughly as the boundary between the northern and southern halves of the country in this period was the great Changjiang, a formidable barrier in those days, which was China's artery not only in irrigating the rich agricultural land all along its basin but also as the main waterway linking east and west. In its upper reaches around the Three Gorges where a great dam is at present being built, the river flowed fast over rock-strewn rapids which made it extremely dangerous for navigation. Scenically very beautiful, it has inspired numerous poets and attracted tourists, but it was also the scene of a tragic living saga of human endurance. The rapidity of the current meant that journeys upstream could be made only under tow, and the tow-paths on the steep rocky shores were so perilous that the towing could be entrusted to no beast but men. The following poem *Ballad of the Three Gorges* is a memorial to the sweat of these straining men, countless of whom, by one false step, would have been swept off their feet to a watery death.

Ballad of the Three Gorges

Anonymous

At dawn we start from Brown Ox Hill.
At dusk we stop at Brown Ox Hill.
Three days and nights we toiled and slept,
But Brown Ox Hill seems standing still.

Once across the River, the scene changed. The Han ruling class who fled the chaos in the north found in the south wide tracts of rich arable land awaiting exploitation, for although the region had already for centuries been under Han rule, it was not given its due importance and hence was economically still underdeveloped. The fugitives brought with them the know-how of the more advanced civilization of the north, and bringing it now to bear on their new dominion, soon made of it a land of plenty in comparison with the devastated northern half of the country. Unfortunately, they also brought with them the shortcomings that lost them their empire in the first place, namely the corruption, the extravagance, the mismanagement, and the incessant dissensions among themselves. The new found riches were squandered on luxuries for the ruling class instead of being used to strengthen the nation and eventually to regain the lost regions from the invaders. Besides, with their abject failure in the north, a new sense of passivity and fatalism had crept in among the ruling class, with the help of indigenous Taoism and of Buddhism imported from the west. Thus one often reads recorded in history some emperor or other giving a sermon on Buddhist scriptures, or a minister-general holding forth on the mysteries of Tao, when invaders were literally hammering on the gates. Besides, corruption was such that one really needed to be thick-skinned to advance in officialdom, and there grew up a fashion of voluntary retirement from public life into a life of leisure, which was seen not as irresponsibility but as a mark of high-mindedness. Naturally, such a view, which persists even today, also found its way into poetry, of which an archtypical representative is **Tao Yuanming** (365–427 A.D.), the dominant literary figure of the Eastern Jin dynasty, an example of whose work follows.

Wordless

Tao Yuanming

'Mongst people still is pitched my hermitage,
But visitors in carriages I've none.
And how have I so managed it, you ask?
When far in mind, one's everywhere alone.
To pick chrysanthemums beside my fence,
I see Mount Zhongnan shimmer in the sun.
The mountain air is fresh near dawn and dusk,
To which the birds will turn when day is done.
There's deep philosophy to tell in this,
But words are gone before they are begun.

Tao Yuanming was descended from an illustrious family, and though not affluent was at least counted among the landed gentry. Even then, he had difficulties making ends meet and sought several times since his youth to augment his income by joining the imperial service. He served for short periods in various minor posts, the last time as district chief of Pengze when he was already over forty. He lasted only eighty-five days at this post before he was again forced to resign in disgust at the corruption and the need for a constant display of submission to superiors. After that, he lived in retirement with his literary pursuits, and died in poverty some twenty years later.

Tao Yuanming was far from alone in that period in his disillusion with public life. **Bao Zhao** (412–466 A.D.), with a more plebian background, had presumably an even harder time, as recorded in the poem cited next. This was written in his youth when he still had the reckless energy to rebel. Circumstances however soon taught him to adapt, though fortunately without entirely damping his spirits. He passed the rest of his life at various minor official posts until he was killed serving the wrong prince in one of the many struggles for succession.

The trouble was that in the monolithic society of old imperial China, there was but one gentlemanly profession: imperial service, which was effectively the only pathway to wealth, consequence and honour, but in the decadent period that Tao Yuanming and Bao Zhao lived in, higher ranks were accessible only to the well-connected. Although both had achieved early a reputation for literature, this did not count for very much, for in those days writing poetry was not a profession but a pursuit that every educated person engaged in, some better than others and if so applauded, but was no substitute for a good lineage or good connections.

Resignation

Bao Zhao

I sat at table, but nought could I partake, -
Struck column with my sword, but only into sighs to break.
Ah, dash it all! It's not so many years a man can live,
To spend them thus with drooping wings, in deep obeisance quake?
I'll chuck this wretched job and go back home.
Could do with a rest, for goodness' sake!
At dawn took leave of parents, dusk return
To parents's side, a little cheer to fake,
To idly play with child, while watching wife
At weaving-frame her daily produce make.
Through history, sage and able men stayed common folk and poor,
Then let alone plain fellows like me none would notice take!

For this reason, the dilemma that Tao Yuanming and Bao Zhao faced was well-nigh universal among those of common origin in that period. This is not to say that there were no fine poets in the ruling class, nor that their life would necessarily be easy. For example, **Yu Xin** (513–581 A.D.) was highly born and since his youth had mingled with princes at the Liang imperial court and made himself a name both as a courtier and as a poet already at a young age. The dynasty did not last long, however, and it fell when Yu Xin was on a mission to the northern kingdom of (Latter) Wei where, because of his literary fame, he was made much of. When Liang fell, he was kept in Wei against his will though with a high honorary title at court, and so passed the rest of his life sadly in exile. It was then that he wrote the poetry for which he is now remembered.

On the whole, however, in spite of all the attention that the ruling class of the Southern Dynasties had paid to literature and their fine sophistication, their output have somehow to the present reader a hollow ring to it. They wrote about the beauty of nature, the pleasures of the retired life, and the purity of Taoist and Buddhist ideals, but very often they were among the same people at court who beggared the nation with their extravagance and engaged in brutal intrigues which periodically decimated their own ranks and often also the population.

Technically, much progress was made, such as the introduction, by Bao Zhao in particular, of "long lines" with seven characters or more, which greatly enhanced both the tonal quality and the power of delivery, especially when judicially mixed with traditional lines of five characters each, as in *Resignation* cited above. Also, the engagement of the educated class had greatly broadened the scope and heightened the sophistication of poetry, giving us a foretaste of the great blossoming which is to happen later in the Tang dynasty.

A String my Zither Broke

Yu Xin

In Jinling once I had a home near county-hall.
To Chang'an married since, now here they me instal.
With tear-filled eyes, I far horizons daily search -
But where, in which direction, does my old home fall?
This dust from foreign climes — ah when will it subside?
And when, the Han full moon, will it again befall?
Because I found you knew this tune, all unawares,
A string my zither broke, and broke my heart withal.

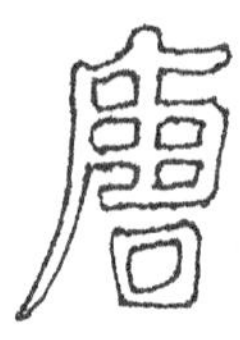

The Tang Dynasty

While the ruling class in the south sank deeper into decadence, that in the north descended from the nomadic invaders had gradually lost their initial savagery, with some indeed among them becoming quite wise in statecraft. One in particular, Yuwen Tai, nominal minister though de facto ruler of (Latter) Wei with capital near present-day Xi'an, built up a kingdom with increasing influence which, after a further couple of changes in dynasties, ended up with Sui, whose founder Yang Jian finally succeeded in 590 A.D. to reunite the whole country. At its height the Sui dynasty almost rivalled Han in the size of population and area under its control. Unfortunately, peace and unity lasted barely two decades at the end of which the country was again plunged into chaos soon after the founder's succession by his profligate son. The country had to go through yet another devastating period of war and turbulence before settling down and reuniting under Tang in 624 A.D.

By all accounts, Li Shimin, the architect of the Tang dynasty, was something out of the ordinary. He was the son of a provincial governor when the Sui dynasty started to break up, and it was he, then only seventeen, who urged his father to make a bid for empire. From an early age, he showed unusual foresight in collecting about himself not only intrepid warriors to help him win an empire but also wise thinkers and able administrators to help him form and manage it afterwards. Himself a brilliant general and expert bowman who habitually charged at the head of his army, he planned and executed one successful campaign after another to put his father Li Yuan on the throne, whose own mediocrity and small-mindedness tended rather to hinder than to assist in the process. And when Li Yuan, after a decade as Emperor, finally abdicated to leave Shimin in charge, China seemed truly to have entered a new golden age.

Wise in deliberations and brave in acts, Shimin as Emperor Taizong was ably seconded by the men he had gathered around himself. His old comrades-in-arms he honoured but barred from government, to save them, as he said, from possible misconduct through conflicts of interest, and himself from the ingratitude of inflicting then the necessary punishment. Considerate of the people's plight, and moderate in his own tastes, he and his assistants lowered taxes and levies, carefully nurturing the country back from the damage of the civil war following the fall of Sui, which left the country with barely a third of Sui's population. In relation with neighbouring nations, he wielded judiciously the carrot and the stick. First demonstrating his military power in some well-conducted campaigns, he invited next their chieftains to his capital Chang'an where he entertained them lavishly and showered them with gifts, and these latter, amazed at Tang's prosperity, which in their grassland homes they could not even dream of, worshipped him as the Great Khan from Heaven. In return for some guarantee of peace and justice on their subjugation, he demanded no taxes, just military service when needed. This was great benignity compared with the norm then in the region, where whoever happened to be ascendent would treat the conquered nations just as slaves. As a result, the Tang Empire under Shimin extended into the north and west much beyond China's present boundary (see map, Appendix I). Secured then from invaders abroad and blessed with good management at home, the population increased quickly in number and in well-being, and the country in wealth and strength. Indeed, the couple of decades of Li Shimin's reign has gone down in history as one of China's very best periods.

With peace and prosperity, the arts began to flourish, in particular poetry which was to reach in the Tang dynasty a peak never attained before. Li Shimin himself, however, was not to see this in his lifetime, for obviously, with the country just recovering after the fall of Sui, poetry was not the first priority. This was to come only in the reign of his great grandson Li Longji, nearly a century later. But even before that happened, poetry seemed already to have shaken off the decadent tendencies of the Southern Dynasties, and acquired both a greater depth of purpose and breadth of vision.

Chen Zi'ang (661–702 A.D.), with a distinctive fluid style of his own, was an outstanding example of this Early Tang period, although he was actually most active in the reign of Wu Zetian, when the country was officially no longer Tang but Zhou.

Wu Zetian was the unique reigning empress in Chinese history. Both before and after her, other empresses had ruled the country in the name of their sons, but Zetian alone had reigned in her own name and even founded her own dynasty. The daughter of a provincial governor, she was selected for her beauty at the age of fourteen into Taizong's harem. Though occupying there but a lowly position, she managed to gain the surreptitious attention of the crown prince, who soon after ascending to the throne as Gaozong at his father's death, flouted all traditions to make her his empress. She soon had Gaozong eating out of her hands, and long before he died he had already effectively ceded to her the reins of government. After his death, she reigned as regent for a while in the name of some puppets she put on the throne, but then dissolved the Tang dynasty altogether, renaming the country as Zhou with herself as Empress, holding sway against an aristocracy still largely faithful to Taizong's memory, and was finally deposed in a palace coup only at the ripe old age of eighty-two.

Even with her great ability and the asset of a docile husband, it would still be doubtful whether Wu Zetian could have succeeded to the extent she did without the beneficial mixing of cultures and change in social conventions brought about by the nomadic invasion the centuries before. Though still male-dominated, women in Tang enjoyed a greater freedom and a higher social status than, say, in the Han dynasty. Li Shimin's sister, for example, was at one stage a general in active service in command of an army.

The Distant View on Top of Youzhou Tower

Chen Zi'ang

Into the past, I cannot see the ancients,
Nor into future, those who come behind.
My heart is filled with anguish, silent tears course down my face,
When thus alone in all Eternity myself I find.

After Wu Zetian was toppled, the Tang dynasty was restored, with as emperor a former puppet under her thumb. He too, however, soon fell under the dominance of his empress and the dynasty faced yet another dissolution, were it not saved in the nick of time by an audacious coup engineered by the young prince Li Longji, which put his father on the throne and the dynasty back on its feet. Like his great forebear Li Shimin, Longji was but the second son and had by tradition no right to succeed his father, but having achieved so much already and so outshining his brothers in ability, he was named emperor when his father abdicated. Longji's long reign (713–756 A.D.) as Emperor Xuanzong divides into two periods. The first, called Kaiyuan and lasting nearly thirty years, almost rivalled Shimin's reign in fame for exemplary governance. Besides, the country having by then recovered from the chaos from which the dynasty emerged, had for the first time in seven hundred years surpassed in size of population and area under its control achieved in the Han dynasty, and aided further by the advance in technology since then, had entered a period of prosperity unprecedented in Chinese history. Arts and crafts flourished and, facilitated by an ever-widening network of roadways such as the Silk Route, trade expanded with other nations, and to the latter Tang must then have appeared as the epitome of civilization. Even today, China towns all over the world are still called *Tangrenjie* (i.e. Tang People Streets) in Chinese.

Much of this success can be traced to **Li Longji** himself. Aided by his own considerable intelligence and by the able men he had gathered about him, he applied himself to government. Indeed, at one stage, as it would appear in the following poem by him, he even aspired to the traditional Confucian ideal of emulating the saintly emperors of legend in benevolence and austerity.

Worship at Shrine of Master Kong

Li Longji

Who was the Master, what was he, that 'mongst
The great of old we him alone adore?
At Zoushi, where the palace of the King
Of Lu now stands, there was his home before.
A phoenix strayed, a captured unicorn,
Foretold that he neglect in lifetime bore.
So may this worship he but dreamt of once,
Faith in the triumph of his Way restore.

Li Longji, however, had no great aptitude for austerity. Born and bred in luxury, he was prone to extravagance and it was this in the end which proved his undoing. However, he appreciated and patronised the arts, and this helped to make of his reign what one may say the apogee of Chinese poetry, both in the height of achievement and in the breadth of participation on nearly all strata of society.

Within the galaxy of great poets in his reign, two in particular stood out, Li Bai (also often transliterated into English as Li Bo or Li Po) and Du Fu, whose pre-eminence was acknowledged even in their lifetime. For the formal perfection of his poetry, Du Fu, the poet's poet, was known as the Saint *(Shi Sheng)*, but it was the free, seemingly unbridled style of Li Bai, known as the Wandering Immortal *(Shi Xian)*, which most captured the popular admiration. The two were good friends, with Li Bai senior in age by about ten years. They lived in a time of great happenings which gave full rein to their genius, and supported further by a cast of contemporaries each of whom would have been good enough to be the crown-jewel of a lesser age, they give us a vivid picture of the period.

In the first half of Li Longji's reign, with the country prosperous and at peace, one sees a picture not unlike a lanscape from that period where everything is painted gold and green. In the following poem *Taishan*, for example, we find even **Du Fu** (712–770 A.D.), who usually paints in sombre colours, in an expansive mood.

Taishan Peak

Du Fu

Ah, how describe this Lord of Mountain Peaks?
It lies where Qi-Lu's verdure lingers still;
Whose dark and sunny sides part night from day;
Whose airs and graces Nature's best distil.
Fresh clouds are born each morning from its depths,
To which, in homing flocks, birds nightly spill.
And soon I aim to reach the very top;
Then all around me, small seems every hill.

Taishan is in modern Shantong province, the territory of the ancient kingdoms Qi and Lu dating from the Spring and Autumn period.

To **Li Bai** (701–762 A.D.), with his more optimistic nature, few things presumably would have seemed impossible or difficult. The scene of the following poem was set in the Three Gorges, that beautiful but treacherous section of the Changjiang which was also the setting for *Ballad of the Three Gorges* cited above. The distance from Baidi to Jiangling is about one thousand two hundred *li,* that is, roughly 600 kilometres. Although the river is fast flowing there, it could well be doubted whether the downstream journey by boat could really be made within a day. But possible or not, the ride would certainly be exhilarating, and the excitement is captured in this popular poem known to most Chinese by heart.

Ironically, this most joyful of poems was written not in the glorious days of Li Longji's acsendance but soon after his fall, when Li Bai himself was on his way to exile for having been drafted into the service of a prince who eventually lost out in the struggle for succession. It was at Baidi that Li Bai received his pardon from the new Emperor, and in the excitement of the moment, and in the hope of the country soon returning to its former greatness and himself to favour at court, this poem was composed. Those hopes, unfortunately, were not realized. Nevertheless, few poems, whenever they were written, could have reflected better the optimism of that earlier golden age.

Down the River to Jiangling

Li Bai

From Baidi wreathed in dawn-tinged clouds I started on my way,
A thousand li to Jiangling, I'll be back within a day.
While endless flows the River, and monkeys high on both banks wail,
Our boat flies lightly past ten thousand layers of mountains grey.

So much then for the optimism of poets belonging to the upper class. But what about the ordinary people; how did they feel? Here is a description of a rural scene in the north of the country at that time, by that master of word pictures **Wang Wei** (701–781 A.D.), famous also as painter and calligrapher, of whom it was said that in every one of his poems is a painting and in every painting a poem. The general impression this gives is one of peace and tranquility, of adequacy, if not affluence, and that life was lived then as it ought to be lived, where ordinary things that really matter, like the old man's anxiety for the shepherd boy, had their rightful place.

Pastorale

Wang Wei

The setting sun slants o'er the village mart,
While cattle, sheep trudge home down narrow lane.
An old man, anxious for his shepherd boy,
Waits by the door, his hands on thorn-staff strain.
Some pheasant calls from far, where wheat is green.
The silkworm sleeps; few mulberry leaves remain.
A farmer, hoe on shoulder, walking past,
Stops by to talk awhile of sun and rain.
The peaceful scene my heart with yearning fills.
How long from this will duty me detain?

And here is another cameo description by **Cui Hao** (704–754 A.D.) of life among the river folk in the south who were mainly traders and merchants plying the waterways near the estuary of the great Changjiang. A more clement climate combined with the fertile soil made of the river valley the granary of ancient China and life there, at least in peace time, must have been quite congenial. It appears that boys and girls had then both the leisure and the freedom to flirt a little when the occasion arose. In any case, they certainly seemed not to be worried about their next meal, or of meeting with sudden violence around the next bend of the river, which is already a great deal to say about an ancient (or perhaps even a modern) society.

Chance Encounter Afloat

Cui Hao

She:

"I wonder, sir, where you might live.
Hengtang is home for little me.
I stop my boat, 's I ought, to ask,
In case that we might neighbours be."

He:

"By Jiujiang's water is my home,
And that's as far 's I ever get.
So we are Changgan people both!
Just, being young, we've never met."

That the country was in good shape did not mean of course that individuals were always satisfied with their lot, and poets in particular are prone to disenchantment with reality. But the stability of the society then at least permitted the disappointed poet the luxury of either retreating into a life of leisure, as depicted in the next poem by Wang Wei, or drowning his sorrow in wine as Li Bai did in the poems following.

Wang Wei is ever elegant in both style and substance. Indeed, if it was thus to be retired after disappointment as described in this poem, then retirement would be a life to be envied rather than deplored.

A Parting Cup

Wang Wei

Dismounting, first to drain a parting cup,
I asked you what you planned to do.
You said, since things had turned out not as wished,
Then back to South Mount's slopes you'd go.
"But ask no more, - let's leave it so,
For white clouds there will never cease to flow."

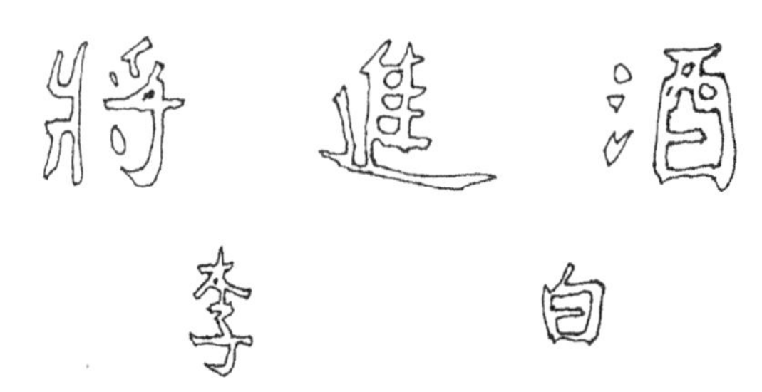

Li Bai's *Tonight We Feast* is an acknowledged masterpiece in the free-flowing style that only he could carry to perfection. Constructed out of lines of varying lengths, ranging from three to ten syllables each, the poem rushes along, wave after wave, like "the Yellow River down from Heaven", unfettered and unstoppable. Indeed, familiar as it is, every recitation of it is such a pleasure that one has the urge to start again from the beginning as soon as one has come to the end. In the present English rendering, an attempt has been made to preserve as much as possible the structure of the poem in rhythm, rhyming pattern, and line lengths, but, minus the tonal structure, much of the genius of the original is unavoidably lost. One can only hope that some of the spirit remains.

The Lord (King) of Chen mentioned in the poem was Cao Zhi, the author of *Beans and the Beanstalk* previously cited.

Tonight We Feast

Li Bai

O see you not my friends the Yellow River down from heaven churns -
A-rushing headlong to the sea and ne'er returns?
And see you not your hair this too clear mirror shows without respite -
Black silk at morn, by evening is turned snow-white?
Just so, when Fortune smiles we must enjoy in full,
And ne'er forsake the bottle when the moon is bright.
Our gifts are Heaven-sent which must their purpose have.
And squandered gold a thousand gone will come again.
Let sheep be slaughtered, oxen slain, for t'night we feast.
At this, one single sitting, three hundred cups I'll drain.
My dear sir Cen,
And you, Dan Sheng.
Pass round the cup,
No slackening!
And pray, lend me your ears my friends
For 'tis to you this song I sing:
The honours of this world are trash, and I but wish
To be forever drunk without awakening.
Through all the ages only drinkers' fame remain;
Of all the rest is silence, be they sage or king.
At Pingle once the Lord of Chen carouse did hold;
Each measure of the wine he drank cost ten in gold.
Why say that you are poor my host? Just let me call
For more, and in this cup I'll show you wealth untold.

Caparisoned steed,
Rich garments gay,
Just call the boy, send him with these the best wine pay,
Then you and I shall wipe the sadness –
from all the history of the world away.

But it is not only in the free unbridled style that Li Bai excels. Here is one in a more formal format with uniformly five characters in each line and a rhyme every other line. It is a format which fits him better, though still drinking, in this pensive mood. Indeed, are human relationships often of much more account than this?

A Flask of Wine Alone among Flowers

Li Bai

A flask of wine among some glorious blooms,
But not a friendly soul can there be found.
Why then, I'll raise my cup to toast the Moon; -
We're three now, with my shadow on the ground.
The Moon, of course, it knows not how to drink; -
My shadow, it just follows me around.
But still, enjoy ourselves in Spring we must,
Though means for doing so may not be sound.
I sing, the Moon, it wanders to and fro; -
I dance, my shadow does a little round.
While yet awake, carouse together, drunk,
We part, for by no warmth or kinship bound,
We just accost each other far across
Deep space, where reigns the silent night profound.

Unfortunately, the golden period of peace and prosperity depicted in the preceding poems lasted only for about thirty years. Li Longji, relaxed and emboldened by the tremendous success of the first half of his reign, became both extravagant and careless. Unlike his great forebear Li Shimin, he won his empire not by military exploits but by a palace coup. He had thus no real knowledge of or ability for military matters, and sometimes sought purely military solutions where there could be none. Nor did he fully appreciate the delicate balance of peaceful and warlike measures that Shimin had so thoughtfully put in place to guard the Empire's frontiers. Hence, for example, he permitted, on the one hand, his commanders to extort more and more taxes in addition to military service from the subjugated neighbouring nations, and even denied the victims justice when they appealed to him. On the other, he allowed the state enterprise for breeding horses, which at one stage had as many as three-quarter million horses in stock, to dwinle to less than half that size, leaving the country dependent on foreign supply for an essential of military power in his days. Thus gradually, he lost among his neighbours both the means for retaining the faith of the majority and that for daunting the ambitious. His fortunes were not helped by the sudden emergence of Tibet as a great power, nor by the expansion of Islam in the west, at the hands of the latter his army once suffered a resounding defeat. Meanwhile, the cost of these campaigns as well as his own extravagance were draining the resources of his empire, which though vast were beginning to feel the strain. As a result, the people suffered, and Du Fu, though loyal imperialist to the core as tradition at that time demanded, could not with his poet's sensisitivity but be pained as he saw a forced gang of soldiers marched off to a war from which they most probably would never return.

Chariots Rumble

Du Fu

Chariots rumble,
Horses neigh.
Each carrying bow and arrows, they're hurried on their way.
Father, mother, wife and child run after them.
The dust's so thick that Xianyang Bridge is hidd'n away.
They tear their clothes, they stamp their feet and bar the road.
Their cries reach up to Heav'n. – If Heav'n could hear today!
A passer-by asks some among the marching band.
They only know that it's conscription, a war's at hand:
"At fifteen, one gets sent off north to guard the River;
At forty, west to work the frontier's virgin land.
When first called-up, one leaves, the darling of the village;
Discharged grey-haired, still stationed by the desert sand.
In frontier lands, blood like water flowing free.
Still: 'Open up the frontier' 's the Emperor's decree.
And know you not, sir,-
In all two hundred counties east beyond the range,
The farmland, mile on mile, o'ergrown by brambles be?
E'en where some woman's strong enough to work the plough,
There's really nothing growing in the field you see.
Besides, the enemy's so tough and awful strong,
Be routed, chased around like dogs and chickens, we.
Since you kind interest take,
We dare complaint to make."

For instance, now, when winter's come,
They'll still not let us soldiers home.
The district-chief'll ask for taxes,
But where to get the gold, wherefrom?
It's ill to be born a man,
Much better to be born a lass.
A lass may still be married to a neighbour;
A man but left to rot like winter grass.
Do not you see,
*By the Green Sea's shore,**
For aeons, neglected bones are scattered more and more?
The ghosts of newly dead complain, the old but weep,
B'neath louring skies on dank earth where wild winds roar."

* The Green Sea *Qinghai* is (or at least was) a large salt-water lake around which numerous strenuous battles were fought between Tang and the then powerful Tibetan empire which at one later stage even occupied briefly the Tang capital Chang'an.

Li Bai, less traditionalist in outlook, and having been brought up on the northwest frontiers, had had more contact and hence presumably also more sympathy for non-Han ethnic peoples. He took, it seems, an even dimmer view of the situation. He dared not, of course, openly challenge the authority of the Emperor, which even in a poem would have risked accusation of treason. Most of the following poem seems thus to be an exaggerated eulogy to the prowess and invincibility of the Han army followed by a heartless celebration of their victory over the barbarian foe, that is, until the last few lines when in a sudden burst of sarcasm, he showed his real opinion and where some of his sympathy lay. The phrases in quotes are a parody of a poem by Liu Bang, the founding emperor of the Han dynasty, written when he first won his empire.

The term "Hu" was used originally to denote either the non-Han nations in the north-east *(Donghu)* or those in the west with deep-set eyes and sharp (European) features, but by Li Bai's time, it covered basically any "foreigner".

Actually, by the Tang dynasty, it almost made no sense any more to speak ethnically of Hans and non-Hans, given the great ethnic mixing by intermarriage during the preceding period. Even Li Shimin, the great Han emperor, had a mother with a "foreign" (that is, non-Han) surname and was thus, it seems, at most half-Han by blood. Art from that period shows strong nomadic influence and often depicts individuals with distinctive non-Han features.

Hu is Emptied

Li Bai

The frosty wind has withered th' desert grass.
A host is gathered at the Yumen Pass, -
Commander Huo, three hundred thousand men,
And hordes of prancing horses formed en masse.
Though sheathed, their brilliant swords still shine from far;
Each arrow at their waist a shooting star.
O'er snow, in gilded mail, our troops advance; -
Hu's shafts of no more harm than sand-grains are.
Formations that to foes the fear convey,
Of dragons, tigers, 'nd other beasts of prey.
They say that Venus passed behind the Moon;
An omen that the foe be vanquished soon -
Be vanquished soon.
Their star is dead!
We'll drink their blood and on their entrails tread;
And we shall hang them high up in the sky;
Or by the roadside leave their bodies lie.
Hu is emptied;
Han is great!
Three thousand years Your Majesty will reign in state.
Just sing: "O clouds, how ye are driven by the wind!"
What need indeed for "stalwarts -

to guard the empire's every gate"?

Meanwhile, the drain on the country's resources continued and reached a point when the framework which held it together could no longer withstand the strain. The Emperor's most trusted general guarding the northern frontiers took advantage of the situation and broke into open revolt, casting the whole country into chaos. Rebel forces even overran the capital Chang'an and drove the Emperor himself into exile to the south in what is present-day Sichuan. Although this rebellion was contained and the capital recovered after a few years, it left the country in devastation. Du Fu has left us in his poems a picture of what the people then suffered much more vivid and powerful than any historical record can convey. He himself, like many others, was uprooted and driven by the upheaval to wander all over seeking livelihood for himself and his immediate family, and was reduced at one stage to gathering wild acorns and roots for sustenance. Friends and siblings were scattered, some never to be seen again. The following poem offers but a glimpse of the general desolation.

Autumn Night in Frontier Land

Du Fu

A wild-goose's scream brings fall to frontier land
Where rumbles of the battle-drum are rife.
The moon is not as bright as ours at home,
Though frost as white will by tonight arrive.
My brothers, they, all scattered are, and there's
No home to ask if dead or still alive.
The letters one can send are not received,
While yet no end is seen to war and strife.

These upheavals so destroyed the fabric of the Tang dynasty that it never fully recovered. Although the dynasty lasted formally for another hundred and fifty years, it was but a shadow of itself in its heyday. It never regained even full control of China proper, let alone the far-reaching influence in foreign lands it once enjoyed. The emperor of the day controlled only the provinces close to home, while the others were ruled more or less autonomously by the provincial governors, who though formally still acknowledging the emperor as overlord, levied their own taxes, managed their own economies, appointed their own successors and fought with one another to increase their sphere of influence. And of course, in all these manoeuvres, the people were the pawns just to be taxed and sent off to battle, unless they were lucky enough to be ruled by a conscientious governor, and such indeed existed, with some of their interest at heart. In the following poem the picture painted by **Li Yi** (748–827 A.D.), active a generation after Du Fu, is still a dark and pessimistic one, no longer in the gold-green colours so typical of the Tang dynasty at its peak.

Brief Encounter with a Cousin

Li Yi

Upheavals great for years have parted us.
This once we met since both to manhood grown,
As strangers greeted till the name you gave
Recalled the face I once so well have known.
Ah, fields have turned to marshes since last we met!
We talked till silenced by the night-bell's tone.
By dawn we'll each again be on our way,
Between us mountains by the wild wind blown.

Although the fortunes of the Tang dynasty had receded, the flowering of poetry it engendered remained, lasting all through the dynasty and even beyond. Indeed, the love for poetry had by then permeated the whole society regardless of rank or location. The emperor wrote poems, as seen in the example by Li Longji cited above, but so did his jester at court ...

One day, it seems, Longji ordered his jester **Huang Fanchuo** to make a joke at Liu Wenshu's expense. Now Wenshu, a courtier who, because of the special location of his beard beneath his chin, was often said to have a face resembling an ape's, offered the jester gold if he would not say so in his response. The following poem, as the story goes, was the result.

Poor Old Liu Wenshu

Huang Fanchuo

O dear o dear! O poor old Liu Wenshu!
His beard and chin in separate places grew.
His face perhaps may not look like an ape's.
An ape, though, looks exactly like Wenshu.

... and so did the women in his palace. ...

Du Qiuniang at fifteen belonged to the household of some provincial governor, after whose death as a rebel, she was confiscated into the imperial palace. In later life (c. 821 A.D.) she acquired a more responsible position, but in her youth, she would be like many other women kept in the households of noblemen or the imperial palace, whose only function was to give pleasure to those who owned them. In her case, it seems, this included the composition of lyrics to abet and justify this pleasure, and indeed, no one reading this poem can say she had not done so admirably.

Pick your Flowers when Full in Bloom

Du Qiuniang

O treasure not, my lord, rich garments gay,
But treasure rather, sir, each youthful day.
And pick your flowers when they are full in bloom,
Not wait to pick them faded all away.

... And so did the burgher in the rich balmy south, ...

This poem is on a then popular and well-worn theme of women living in seclusion missing their men who had been posted to the frontiers, though viewed from a rather unusual angle. The only thing known about the author, a man, seems to be that he lived in Lin'an, that is, present-day Hangzhou, that beautiful scenic city by a lake which is a great tourist attraction even today. In peacetime, it must have been a pleasant place to live, being right in the midst of the so-called region of fish and rice, which is the Chinese equivalent of a land of milk and honey. There, no doubt, one can afford even to pick a quarrel with the nightingale.

O Shoo away that Nightingale

Jin Changxu

O shoo away that nightingale,
And stop her singing in the tree.
Her singing woke me from my dream
Just when my love I hoped to see.

... as did the serf on the land in the rugged frontier region of the far north-west.

Kan Man Er, active in the reign of Emperor Xianzong (806–820 A.D.), belonged to the Uygur (Uighur) ethnic group in what is now Xinjiang in north-western China. Three of his poems, dated 816 A.D. and hand-written in Han (Chinese) characters, of which this is one, were unearthed from the ruins of an ancient city in 1959 and are now preserved in the Xinjiang Uygur Autonomous Region Museum. The poet seemed not to have much exposure to the then current poetic tradition, or if he had, he was deliberately flouting it, for the poem adheres to none of the accepted rules. It has lines of varying lengths — indeed, a very unusual mix with three, four, five and seven characters each — the rhyming pattern, if any, is irregular, and the language unrefined. Yet, by sheer force of passion, the poem hits us with enormous power. For this reason, in rendering this poem, the translator has departed from his usual practice of attempting to imitate the structure of the original for, not sharing the original poet's strength of feeling, it would have been futile to try reproducing a similar effect with the same meagre tools.

Wolves of the Manor

Kan Man Er

Those wolves of the Manor are an evil brood.
They rob our provisions and suck our blood.
Before the cloth can leave the loom or grains the land,
They are all already taken from our hand.
But soon, ye gods, will come a day,
When Heaven will crumble and Earth will crack these wolves to slay.
Then off, my friends, the clouds will fly,
And you and I will see again the clear blue sky.

And they wrote poems on all sorts of themes; historical events, given the Chinese fondness for history, were particular favourites. They would write a poem as a comment on a historical figure...

The following poem by **Du Mu** (803–853 A.D.) on Shihuangdi is an example. When Ying Zheng founded the Qin empire in 221 B.C. after subjugating the other Warring States, he thought it was for perpetuity. That was why he called himself Shihuangdi, meaning the First Emperor of a line stretching on to infinity. But, as already noted, his empire lasted only fifteen years, toppled soon after his death by a popular rebellion led by Xiang Yu, who sacked and burned his capital Xianyang near present-day Xi'an. Legend has it that his vast tomb itself, never yet unearthed, was burnt, apparently by accident by a shepherd looking for a lost lamb. Both Xiang Yu and Liu Bang, who founded the Han dynasty after him, were of common origin and were said to have watched the procession among the crowds when Shihuangdi passed by on his grand tour of the empire after his accession to the imperial throne.

The original of the present poem has actually only the standard lines of seven characters each, the unusually long lines in the translation being just the fault of the translator who has found it hard otherwise to cram in all the historical facts alluded to which would be familiar to the Chinese reader.

The First Emperor

Du Mu

Across his world-dominion, the First of Emperors
now travels far and wide.
Among the crowds along the way, both Liu and Xiang stand watch,
and by their turns abide.
No easy task that, levelling by edge of sword alone all Land
o'er-roofed by Heaven!
But surely, it was not just meant for these two wretched fellows
watching by the side?

It seems, milord, that in the end the people aren't the Fool,
but you are. - Does it not?
Those palaces and long great walls that, like a prison, you have built
to keep you safe,
From very top to far, far underground has crept a shepherd's fire
and burnt the lot,
Turned all to ash, before your pompous-buried body
has had proper time to rot.

... or to describe and embellish a historical event to lend it greater poignancy, as in the *Bronze Man's Farewell* below. ...

Maoling was the burial mount of the Han emperor Wudi whose long reign (140–84 B.C.) was distinguished by numerous large-scale campaigns against the non-Han ethnic nations of the north, west and south. Although these campaigns were largely succsssful in the sense both of curtailing the incursions by these nations and of vastly extending Han influence, they bled the empire of its resources, both human and material, and left it considerably weakened. Being by nature as well as by force of circumstances a little of a megalomaniac, Wudi found himself near the end of his life unable to face the fact that like all others he would die, and tried by various magical means to gain immortality. Among the many suggestions made, one was that he should collect the moisture from Heaven which had never yet touched and been polluted by the Earth, and drink it with powdered jade. For the purpose of collecting this *ganlu* (sweet dew), the essence of the moon, he caused a Bronze Man to be made, said to be two hundred cubits in height and holding in its hand a tray for the dew. Unfortunately, this magic also failed. Some three hundred years later, an emperor of the Wei dynasty descended from Cao Cao had the idea of appropriating what Wudi had missed and sent for the statue to be moved to his own capital in the east. They got it as far as the mountains to the east of present day Xi'an, but because of its enormous weight, could not move it any further. What happened to the statue afterwards, we do not know.

The poet **Li He** (790–816 A.D.), an acknowledged prodigy who unfortunately died young, had an unsually vivid imagery among Chinese poets.

The Bronze Man's Farewell

Li He

The gallant lord of Maoling now with west winds rides.
What horses neighing heard at night the morning hides.
Late blossoms droop where once were painted rails. O' er all
His six and thirty palaces green moss presides.

From far, a lackey of the latest king now sped,
Against eye-piercing winds by which that East Pass's fed,
To fetch the lunar essence which once was Han's! Let fall
Clear tears, for glories past, like drops of liquid lead.

But withered flowers in view, he left, dew-tray in hold.
If Heaven could feel, then even Heaven would grow old.
Weicheng is left behind, the gurgling River fades.
Alone then with the moon in desolation cold.

... Or they would write a poem to tell a story, or to illustrate a fable. ...

A Lost Paradise by Wang Wei which follows was based on a fable by Tao Yuanming, the recluse of *Wordless* cited above. The original reflected the hankering of people in general, and of Tao Yuanming in particular, in those unsettled times for a peaceful life. Wang Wei wrote this, however, when he was only nineteen, and right in the midst of the Tang Golden Age. It is thus interesting to compare the two versions; what was an urgent longing in Tao Yuanming's version became transformed here in Wang Wei's into an idyllic tale.

A Lost Paradise

Wang Wei

A fishing boat among green hills one lovely day in spring;
Along the stream, peach groves in blossom close to both banks cling.
So rapt in wonder at the sight, one loses sense of time.
Ah, how foretell this ancient channel would new tidings bring?

At first but narrow is the passage in the mountainside.
Then all-a-sudden down on level plain it opens wide.
From far, one sees there just a clump of tall cloud-piercing trees,
But near, some thousand homes in flowers and bamboo bushes hide.
The visitor is first, and first a Han-style name to bear;
By Qin out-moded customs still the local folks abide.

From mundane world away they found a new society,
At this secluded spot to lead a life of harmony.
Beneath the pines by moonlight, silence reigns among their homes;
When sun comes out, their dogs and fowls make loud cacophony.

The village quickly gather at the news, excitement spreads,
To ask about their homes, or what is now there in their steads.
At dawn, the village stir to sweep bright petals from their doors.
Their fishing boats return when dusk light haze on water sheds.

To flee disturbance they have left the haunts of Man behind,
And ne'er return as happiness in fairyland they find.

To world at large there are but drifting clouds o'er bright green hills.
Within, no heed is paid to all the rest of humankind.

Although no doubt in finding this that Fortune had a hand,
Cannot resist the urge to see once more the native land,
To bid farewell to those at home, and then return for good
Whatever mountains high or waters deep should 'gainst one stand.
He thought he knew the way, but when he comes again next spring
The hills seem changed in subtle ways one cannot understand.

It's all one needs, one thought, the depths of mountains first to gain,
Then make some turns along the stream, one'd paradise attain.
Come spring, one sees all round but these peach-blossom sprinkled brooks.
No longer can the way to paradise be found again.

... They would write poems, of course, on eternal themes such as romance...

Bai Juyi (772–846 A.D.) was very popular already in his lifetime, not only because he was a fine poet but also because he chose to write in a simple language that, it was said, even an uneducated old woman could understand. He is best known today for his two long narative poems which most Chinese scholars can recite by heart. One tells the romanticised story of the beautiful Yang Guifei, the favourite of the emperor Li Longji, whose love for her was blamed for much of the ills which befell the country in the latter half of his reign. The other tells of the poet's chance encounter while in exile in the south with a *pipa* (a string instrument) player and of her performance on a moonlit night on the River. The first runs to 120 lines and the second to 88, which lengths, for reasons explained in the Introduction, are rare in Chinese poetry. He excelled, however, also in the extremely short, two examples of which are included, the first here in the form of a riddle.

Like Flower, but Not a Flower

Bai Juyi

Like flower, but not a flower.
Like mist, though not a mist.
Midnight, it'd come,
By dawn, desist.
And when it comes, so like a dream in spring, it stays but for a little while.
And then it's gone, like vanished morning clouds, as if it never did exist.

... or on a mother's love, ...

A close contemporary of Bai Juyi of the so-called Middle Tang period, **Meng Jiao** (751–814 A.D.) wrote however in a very different and much less approachable style, which was once described by Su Shi, a star of the Song dynasty whom we shall meet later, as "like eating small fish, –what one gets is often not worth the trouble of getting it". Nevertheless, these famous lines of his, especially the last two, must have brought tears of remembrance to many a Chinese traveller.

Mother

Meng Jiao

The quivering thread in loving mother's hand
Are clothes he wears still as he wends his way.
With every hurried stitch at parting made,
Came all her fears that might return delay.
In what way can an inch-lengthed blade of grass
The radiant warmth it took from Spring repay?

... but they would write poems also for the most mundane purposes in everyday life, to issue an invitation, for example, ...

In those days, newly brewed wine (from grain) often had little specks of lees floating on top (called green ants) and was to be drunk warmed; hence the little red stove. The invitation was issued to a certain Mr. Liu, and although it was not recorded whether the invitation was accepted or not, we can safely assume that it was, for who could have declined an invitation cast in such alluring terms?

Invitation

Bai Juyi

Some new green-speckled wine,
Red earthen stove aglow.
It looks like snow this ev'ning.
Canst drink a cup, or no?

... or to answer a letter, ...

Li Shangyin (812–858 A.D.), an outstanding poet of the Late Tang period, is perhaps most famous for a special style in a romantic vein that he often adopted, of which we shall later find an exmaple entitled, typically, *Untitled.* But to this he was by no means restricted, as clearly shown by *Bashan's Rain* cited next and *Sunset* soon to follow.

Bashan's Rain

Li Shangyin

You asked the date of my return. There's no date still.
Here rains at night the autumn ponds to brimming fill.
Ah when, by westward window we a candle trim,
And speak of times when rains at night o'er Bashan spill?

... or even just as a word-game to challenge one another at parties.

This particular format was apparently quite popular at one stage. One starts first with a line with only one character which sets the rhyme and serves also as the title. Then one follows with a sequence of line-couples increasing in length by one character each time till finishing with a couple with seven characters each, rhyming at every other line all along. Obviously, the result is often a little strained.

It is not entirely fair to **Liu Yuxi** (772–842 A.D.) to cite only this example from him, for he was a significant poet of the Middle Tang period with at one stage even a role in the central government, besides being a close contemporary and friend of Bai Juyi, already cited, to whose work he was often linked. Interestingly, modern research suggests a non-Han origin for both Liu Yuxi and Bai Juyi, these two outstanding exponents of the Middle Kingdom culture, probably without their being conscious or even aware of it themselves. It appears that Liu was descended from the Xiongnu immigrants who opted to be Han subjects when their own empire split near the end of the Han dynasty, as already mentioned, while Bai was descended from Qiuci, a small nation once flourishing in present-day Xinjiang. This just serves as another reminder that by the Tang dynasty, the so-called Han nation was already an inextricable mix, and the Middle Kingdom culture a product of this mixture to which many nations had contributed.

Water

Liu Yuxi

Water -
Most beautiful,
The clearest ever.
Though small as in a cup,
Or vast as in a river,
And whether flowing or standing still,
It is to Man and Life a giver.
Akin in purity to Tao and souls;
Compared to friendship of the true, none better.
Even now, though you above the Golden Valley soar
While I inside a nook beneath the Stone Gorge linger,
Yet still from far our hearts converse in love like running streams.
In harmony through day and night, they will never cease to whisper.

Feeding this popularity, or perhaps because of it, poetical style itself made great advances in the Tang dynasty. Not only were all the older forms amply exploited, but new ones were developed. Poems of that period are traditionally classified into four main categories: the ancient style (*gushi*) and the (ancient) lyrical style (*yuefu*), plus the two "later forms" *lushi* and *jueju.* Each of the four categories are further divided into two subcategories according to the line length, five or seven chracters in each line. This does not mean, of course, that other forms were not tried, but those listed above were the most common. In any case, in the first two categories of so-called ancient styles, basically anything goes. Lines can be of mixed lengths ranging from three to seven and sometimes even to ten or more characters, only with five and seven being the most common. The rhyming pattern also can be varied at will, with rhyming at alternate lines being the most common, and the rhyme can be carried all through a piece or changed several times along the way. The poet has thus full freedom to develop his own style and means of expression. Both *Tonight We Feast* of Li Bai and *Chariots Rumble* of Du Fu cited above are classic examples. In contrast, the so-called later forms are much more restrictive. The *jueju,* for example, has exactly four lines of equal length, each with five or seven characters, rhyming at alternate lines, with a rhyme at the first line optional. There being little space here for development, especially for the five character version, the art is to catch a scene or sentiment within its narrow compass. The *Invitation* by Bai Juyi quoted above is a good example. Here is another by Li Shangyin, which everyone past middle age will appreciate.

Sunset

Li Shangyin

T'wards evening, my mind not all at ease,
I drove my way on to the old plateau.
The setting sun feels infinitely good,
Though only, 'tis now near the twilight glow.

Li Bai's *Down the River to Jiangling* above is an example of the seven character version of the *jueju*, and here is another by him. We should not be duped, though, by what it says, for Li Bai was really no recluse by nature and only chose to live among green hills when the fancy took him. The peach-blossoms are an allusion to Tao Yuanming's fable retold by Wang Wei above in *A Lost Paradise*, a recurrent theme in Chinese literature still alluded to even in everyday life today.

Green Hills

Li Bai

They ask me why I choose to live among green hills.
I smile, not answer. Peace my heart with pleasure fills,
For here's a different world from that where humans live,
Where stream, peach-blossom decked, e'er t'wards the distance spills.

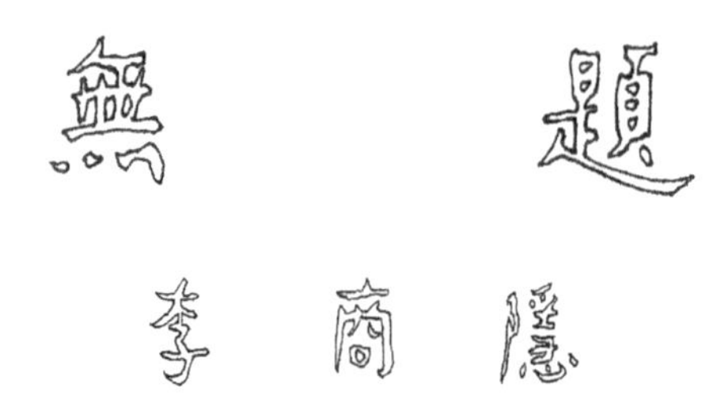

The *lushi,* even more restrictive, must have eight lines of equal length with five or seven characters each, rhyming alternately, with the same rhyme all the way through. What is particular, however, is that the third and fourth lines, as well as the fifth and sixth lines must form *couplets,* a feature perhaps unique to the Chinese language which requires some explanation. As noted in the Introduction, Chinese characters have each a tone. The eight different tones they can have are divided into two classes: the *ping* (level) tones in one class and the *ze* (slanting) class comprising all the rest. An ideal couplet is such that the corresponding characters in the two lines should have tones belonging to opposite classes; thus if the third character of one line is *ping*, then the third character of the other line should be *ze.* Furthermore, corresponding characters in the two lines should belong to the same part of speech; thus if the fourth character of one line is a noun, then so must also be the fourth character of the other line. There can be some deviations, but not much. These requirements may seem excessively restrictive but are in fact not so. The search for a pleasing musical quality already leads rather naturally to the required tonal prescriptions while the relatively uniform syntax of Chinese sentences lends itself quite easily to satisfying the second requirement. Indeed, whether in a *lushi* or not, a poet would often find himself writing couplets in any case. And in the hands of an expert, a couplet can come across with extraordinary force or elegance. For example, the third and fourth lines of the following *lushi* actually read as follows:

Spring - silkworm - meets - death - silk - then - finished
Waxen - candle - turns - ash - tears - only - dried.

which in the original Chinese is quite exquisitely elegant. In English, however, a couplet strictly rendered would seem stilted; I have therfore avoided doing so.

Untitled

Li Shangyin

It's hard when meeting, harder still when parting time is nigh,
For when the breath of spring begins to falter, flowers will die.
*Till death, the silkworm winds itself in endless bonds of silk,**
And only when to ashes burnt, a candle's tears will dry.
The morning glass forebodes this tress will lose its midnight shade.
By moonlight cold, a song at dusk oft turns into a sigh.
It shouldn't be far that happy isle where heavenly maidens dwell,
Go seek for me, you gentle blue-birds wandering free on high.

*Again an exploitation of the coincidence in pronunciation between the characters for "silk" and "thought", as explained in the footnote to *A Length of Silk* above.

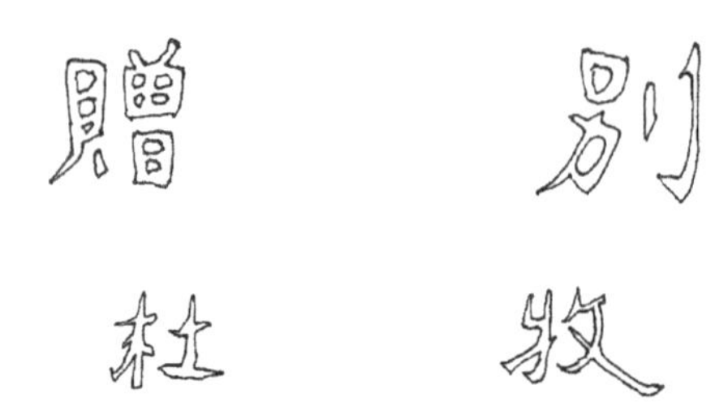

Both Li Shangyin who wrote the *lushi* above and Du Mu who wrote the *jueju* which follows, and both cited already before, belonged to the Late Tang period and both were fine poets, gifted with unusual talent and sensitivity. However, when compared to the poetry of Early Tang and the Tang Golden Age, their work seems somehow to lack the same force and urgency. It is as if the poets of the earlier Tang periods were describing to us a reality they had before their eyes, whereas the Late Tang poets were describing to us a reality which was once removed from them, an idealized or imagined reality which could be but was not truly there. The difference was probably not due just to differences in personality, for Du Mu in particular was very positive in outlook. Though noted as a play-boy in his youth, as indicated in the *jueju* here cited, (which behaviour may nevertheless be interpreted as a positive but vain attempt on his part to seek in pleasure-houses the true love which had then been all but squeezed out of existence by social conventions), throughout his career in later life as a high official both in central government and in the provinces, he was courageous and indefatigable in pointing out the ills of the times. He made repeated representations to the Emperor with concrete suggestions for improvements, which unfortunately, though often applauded, were never carried out. And this was a record of which few of the earlier poets could boast. Rather, it was as if the Late Tang poets, faced with a reality they deplored but could not ameliorate, lacked the heart to tell us in detail about it, and told us instead about a half-fictitious world beyond.

A Parting Gift

Du Mu

That she could feel, I know, but does she feel?
Just from her lips tonight no smile would steal.
This candle seems in pity, though, to take our part
And shed hot tears till dawn, a burning core reveal.

Indeed, in the Late Tang period, the country was in very bad shape. The grand edifice that Li Shimin built had long since crumbled leaving but an empty shell. Fortunately for the dynasty, neighbouring kingdoms such as Tufan (also pronounced as Tubo, i.e. Tibet) in the south and Huihe (Uighur) in the north also saw a decline in vigour, or it would not have even survived. As it was, the dynasty was left to rot by itself, with one weak emperor succeeding another, until there was no longer any semblance of cohesion. The population, oppressed by exploitation and want, rose everywhere in revolt, which though suppressed in some areas soon broke out again somewhere else. The rebel leaders themselves had no vision of what to put in place of the regime they were destroying nor any thought for the well-being of the population whose grievances they exploited. To most of them, the people were just a source of plunder, to be subdued brutally if need be, and sometimes even without any need.

Huang Chao (?–884 A.D.) was one of the more successful rebel leaders. At least at one stage he did briefly attempt to give a human face to his enterprise by restraining his troops from plunder. True to the Tang spirit, he too wrote poems, though not very refined ones. Here is one he wrote after failing twice the examination for imperial service, laying out his ambition, which was indeed realized later when, at the head of a rag-tag army of several hundred thousand, he took the Tang capital Chang'an and drove the reigning emperor into exile. But he seemed to have no idea what to do with the prize he had won except to proclaim himself emperor instead. In a few more years, he himself was defeated and killed. Even then, the Tang emperor had no power to pull the country back together, but managed barely to survive and let it slide further into chaos.

Chrysanthemums

Huang Chao

When the season changed to autumn had,
My chosen flower would bloom, all others dead,
A fragrant host pervade the city through,
And all Chang'an appear in gilt mail clad.

Song and the preceeding Five Dynasties

The Tang dynasty finally ended in 907 A.D., not with a bang but with a whimper, when the most powerful warlord of the moment got tired of paying lip-service to its last representative and made himself the emperor instead of a new dynasty. There followed then a period of some fifty years when five dynasties followed one another in rapid succession. None of the so-called emperors of these Five Dynasties controlled any more than a few modern provinces' worth of territory at the centre, while the rest of the country was partitioned, at one stage into as many as ten autonomous kingdoms. These continued to jostle for power both with one another and with the reigning "emperor" and war and chaos continued.

And once again, as previously after the fall of Han, the weakening of the power at the centre after Tang's demise prompted invasion by the nomadic peoples of the north and west. The peoples of the north were at that time united under the Liao empire, which was militarily much stronger than any of its contemporary Han states. Indeed, in his bid for the throne, the founding "emperor" of one of the Five Dynasties depended on Liao's aid, for which he had ceded to Liao a large swathe of Han territory including the area around the present capital Beijing, together with all its inhabitants. He had further to submit to the Liao ruler, acknowledging him as the "father emperor" and pay him a huge annual tribute. This arrangement was bitterly resisted, of course, by the people affected, and even by the later dynasties, which provoked Liao into frequent raids further south with devastating effects. What saved the day was often only the difficulty the nomad horsemen had of adapting to the warmer and damper climate.

Surprisingly, however, despite the unenviable state of the nation then, poetry still flourished, though perhaps only among the upper classes. The rulers of the various states, while they lasted, lived in luxury which in many cases included an indulgence in poetry, so that their courts were often filled with poets, some of whom became ministers on no other basis than sharing the same taste for poetry as their masters. Indeed, some of the rulers themselves were highly accomplished poets. However, like that of Late Tang, the poetry of the period had an evasive feel to it as if the poet were deliberately describing only the beautiful, and avoided the unpalatable part of reality.

The following poem by **Wei Zhuang** (836–910 A.D.) is an example of the period. This speaks only of pleasure but there is a hint of a darker truth underneath. Indeed, in the poet's lifetime, he was most famous, and justifiably so, for a long moving poem telling the story of a woman living through those difficult times, which showed that he really cared, and cared deeply, for what the people were then suffering. He became late in life a senior minister in the Shu kingdom in present-day Sichuan, indeed on a much better basis than just his poetry, but earlier on, he had sheltered and travelled extensively in the relative peace of Jiangnan, that is the beautiful scenic region South of the River (Changjiang) around present day Suzhou.

A Moonlike Vision

Wei Zhuang

"Ah, life is good in Jiangnan!" Often have I heard it told.
"If ever travel there, you ought to stay and there grow old."
When showers drum on painted boat, you'd lie
Afloat on water bluer than the sky.
A moonlike vision by the stove would glow,
Her wrists as white as newly fallen snow.
So b'fore you're old, never for home return,
Or else for these, with breaking heart, you'll yearn.

The confusion of the Five Dynasties period finally ended in 979 A.D. when Han China was reunited under the Song dynasty, that is apart from the region ceded to Liao in the north. Among the previously independent kingdoms annexed by Song in this process was South Tang, so named because the ruling family shared the same surname *Li* as the Tang emperors and claimed thereby for itself some of Tang's prestige and legitimacy. The kings of South Tang had ruled for some forty years a rich and quite extensive region in the lower Changjiang basin on both sides of the River and at one stage the kingdom was both reasonally well-governed and at peace with its neighbours, and so was a relative haven in those unsettled times.

Li Yu (937–978 A.D.) was the last of the South Tang rulers (hence is more commonly known as Li *Houzhu,* meaning the Last King). Both Li Yu and his father, known as *Zhongzhu,* the Middle King, were accomplished poets, more adept in composing lyrics than in state affairs, and they surrounded themselves with fellow poets as ministers who had no better knowledge of the world than they themselves. So the original strength of the kingdom inherited by Zhongzhu soon wasted away through mismanagement and extravagance, leaving it in a much weakened state, which Houzhu himself did nothing to improve.

In 975 A.D., South Tang's capital Jinling (present day Nanjing in Jiangsu province) fell, and Li Yu was taken to the Song capital at Kaifeng where he passed the last years of his life. Though given a title and nominal honours, he was kept a virtual prisoner and suffered daily the humiliation of such. It was then, however, that he wrote the poems for which he is now mostly remembered.

A Very Special Taste

Li Yu

Alone to silence, up the western tower, I myself bestow.
Like silver curtain hook, so does the new moon glow.
The falling leaves of one forsaken parasol[*]
Make deeper still the limpid autumn locked up in the court below.
Try cutting it, it's still profuse -
More minding will but more confuse -
Ah, parting's such enduring sorrow! -
It leaves behind a very special taste the heart alone could know.

[*] A stately Chinese tree with broad leaves like the sycamore.

The poems of Li Yu which have survived are all lyrics of the type known as *ci* of which he was the acknowledged master. As far as is known, *ci* first made its appearance in the Tang dynasty, and to such well-known Tang poets as Li Bai and Bai Juyi are ascribed the first examples. By the Five Dynasties, it had become very popular, and in the Song dynasty it occupied centre stage superceding all the older forms. As already noted, Chinese poetry has always been intimately connected to music, and there was in particular in Han and Tang poetry a type known as *yuefu* which was meant to be sung or recited to musical accompaniment. This new type known as *ci,* however, is special in that each *ci* is written according to a certain template, known as *cipai,* to fit a specific tune or melody. The template specifies not only the number of lines and of characters in each line, but also the rhyming pattern and even the tone (i.e. whether *ping* or *ze*) of a fair number of the individual characters it contains. The lengths of the lines usually vary depending on the template. For example, in the following *ci* by Li Yu the template specifies that there should be 7 lines with respectively 6, 3, 9, 3, 3, 3, and 9 characters, and that there should be a rhyme at the end of every line but that the rhymes for the 4th and 5th lines should be different from the others. The *cipai* for the *Moonlike Vision* by Wei Zhuang cited above, on the other hand, is simpler, requiring only 4 rhyming doublets, with the first having 7 characters, and the rest 5. These rules are not arbitrary but the result of previous attempts to achieve a particular tonal quality and have in fact to be fairly strictly followed or else the resulting *ci* would not sound right. For this reason in Chinese, one speaks of writing or composing a *shi* but one *tian,* that is fills in, a *ci.* To suit his mood or temperament, a poet can choose from a number of existing templates, or else rarely, if he is a musician as well, create a template of his own.

Rouge-Tinted Wine

Li Yu

So once again I see the spring flower go.
"Why dost hurry so?"
"But how withstand the cold of morning rains and night winds' blow?"
No more repine -
Rouge-tinted wine,
Like teardrops' glow.
Man lives but to regret, as rivers to the eastward flow.

The rules for "filling in" a *ci* being so strict, one would have thought that they would inhibit the poet's freedom of expression and stifle all originality. Surprisingly, however, this is not entirely the case, for the best of the genre not only manage to give no impression of being inhibited in any way, but, probably because of their very special tonal quality, achieve an intimacy and intensity of feeling seldom found in older forms, though often perhaps on a smaller, more personal scale. Of these, the work of Li Yu written in captivity after the loss of his kingdom are particularly good examples. Every line, it seems, knocks one on the heart, each echoing but going deeper than the last. Indeed, for this reason, the *ci* of the Song dynasty is probably the most popularly appreciated poetic form among Chinese readers today, in comparison to which the other forms would seem more distant and remote.

As for Li Yu, he succeeded so well in the *ci* which follows in expressing regrets for his lost kingdom that it gave the Song emperor, it was said, the excuse soon afterwards to put him to death.

A few notes on details are in order for appreciation of both the preceding and the following poem. Because of the particular geographical structure of the Chinese landmass, high in the west, sloping towards the east, almost all rivers flow eastwards. Thus, water flowing east is often regarded as part of the natural order. Further, the climate is such that winds normally come from the east in spring and from the west in autumn, so that "east wind" is synonymous with spring and "west wind" autumn. Lastly, there was, so it was said, then in Li Yu's old kingdom a wine actually called "rouge-tinted tears".

East Wind

Li Yu

Spring flowers followed by the autumn moon -
how much longer will it last?
And who can recall the many things
which have happened in the past?
Upstairs, last night,
I felt again the east wind blow.
Then any longing for my once-held kingdom
let only the bright moon know.
My palace carvings of precious stones should still be there,
But the youthful looks I once had - where?
If you ask me how much sorrow my heart does wring,
It's like the east-bound River swelled by the flood of spring.

Not everyone, of course, had Li Yu's unusual emotional experience to convey, which seemed so specially suited to the *ci* medium. Nevertheless, others had successfully constructed with it some veritable gems of beauty, though often on nothing much in particular. One favourite theme, for example, was, as in Li Yu's *Rouge-tinted wine,* the regret for the passage of spring, but now not for the loss of a kingdom or any specified reason. It was just the feeling of regret in the abstract that was depicted, and it is impressive how many artistic twists they were able to endow it with. The following is one example by **Wang Anguo** (1028–1074 A.D.), ...

She's Learned to Make the Pipa* Cry

Wang Anguo

So all the pleading of the nightingale
For spring to stay has been of no avail.
Like silk embroidery o'er the garden spread, -
The flowers brought down last night by rain and gale.

For th' first time she last night has learned to make the pipa cry.
Since dawn, her thoughts have wandered where th' horizon meets the sky.
What good indeed are jewelled casements and fine-painted halls?
The breath of spring will e'er with errant wisps of willows fly.

* String instrument akin to the mandolin.

... and here is another by **Huang Tingjian** (1045–1105 A.D.).

Despite these successes, however, the *ci* format had inherent limitations. First, although the restrictions imposed by the *cipai* or template are hardly felt in the finished products of a master, they are nevertheless a serious constraint and become more noticeable with templates of greater lengths. The earlier templates were usually rather short with fewer than a dozen lines, as in the examples cited above, but as the form evolved, templates were introduced which had twenty-plus or even thirty-plus lines, and for the *ci* composed with these, it can become apparent that some lines were inserted by the poet only to fill the gaps. Secondly, and more seriously, the tunes or melodies on which the templates were based originated mostly from pleasure houses, or were at least made popular there, sung by courtesans. Although in those days pleasure houses were highly sophisticated affairs and courtesans veritable artists of great talent, there were limitations to the subject matter and the range of emotions they explored, and the *ci* composed on these templates, at least in the beginning, seldom ventured beyond those limits.

Where is Spring?

Huang Tingjian

Does anyone know where spring has gone?
It went so quietly and left no trail.
If anyone knew where spring had gone,
We'd call it back here for a longer stay.
But where is spring? Ah who could say,
Unless we ask the nightingale.
In lilting notes it seemed to tell a tale,
Until through roses with the breeze it flew away.

For this reason, it seems that when faced with subject matter outside that limited range, the early Song poet, even though a master in the *ci* format, would often revert to one of the older *shi* forms of Tang or earlier. The following three poems by Su Shi are examples. The first is a *lushi* and the second and third are *jueju,* all with seven characters to each line in the standard Tang tradition; the irregularity and lack of rhymes in the present English rendering of the first one are due just to the translator finding it easier thus to convey the feel of the original.

As the *ci* format developed and increased further in popularity, however, it was applied to essentially all subjects, and most other citations from the Song dynasty here are in this format.

Su Shi (1036–1101 A.D.), perhaps better known by another name **Su Dongpo,** was widely acknowledged even in his lifetime as the unchallenged giant of Song literature. With his father Su Xun, and brother Su Zhe (Ziyou), known together as the three Su's, they dominated Song literature for more than half a century, and to their family many contemporary figures such as Huang Tingjian cited above, even though well-known enough in their own right, would proudly attach themselves as disciples. Su Shi, in particular, in contrast to his father and brother who were mainly essayists, excelled in almost every branch of the arts, from literature to painting and calligraphy. In literature, he was equally accomplished as an essayist and a poet, and as the latter in most of the existing forms. Indeed, his fame as a poet spread even beyond Song boundaries into foreign climes, so much so that when Su Zhe went on an embassy once to the Liao court, he was confronted by one of the ministers there with a quotation from his brother's poems rendered into a foreign tongue.

Wild-goose on a Snow-bound Field

Su Shi

So what is Life,
But a wild-goose lighting on a snow-bound field,
And if on earth it chanced there to have left a print,
What matters more which way its flight -
or to the East, or to the West?
The monk is dead, and in his name a new pagoda built.
Some poems he wrote remain yet on the broken wall.
Remember still the rugged journey that is past? -
The long way, tiredness, and the cries of laden mules?

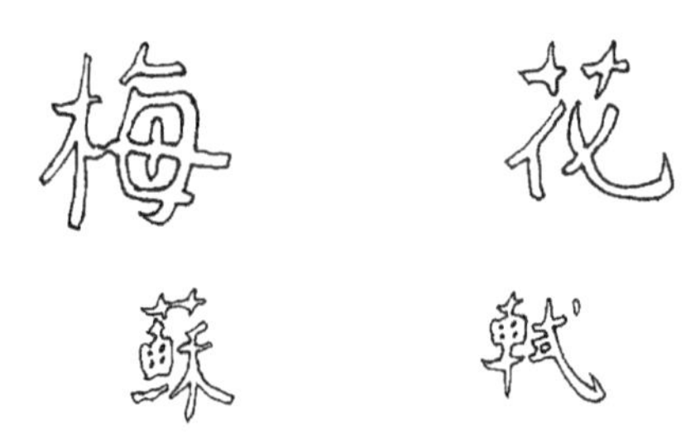

Though recognized early for his ability, Su Shi had a checkered career in the imperial service. After an initial period of steady progress, he was caught up in the political in-fighting at court between the reformist faction headed by the great innovative statesman Wang Anshi and the conservative faction in which Su Shi found himself. Su Shi and Wang Anshi themselves remained on amicable terms, but the opportunistic followers of the statesman did not have his scruples, and when their faction was in power, lost no chance in persecuting their political opponents. Thus, at one stage, Su Shi found himself in prison on some trumped-up charge from which he had little hope of emerging alive. Though ultimately reprieved, he was banished for long years. Only for about eight years, when the conservatives were in power, was Su Shi recalled to fill improtrant positions both in the central government and in the provinces. As provincial governor, he was benevolent, with the people's welfare at heart, paying attention to, and even writing poems on, such practical matters as coal-mining and irrigation. At Hangzhou, there is still a causeway he built named after him. After that, the change in the ruling power again saw Su Shi banished to distant places, finally to as far south as today's Hainan, then desolate and undeveloped, from where he was recalled only at the ripe old age of sixty-three and died on the journey back.

Fortunately, with his rather happy personality tempered by philosophy, and protected by an unassailable reputation, he managed to survive all these reverses with almost no trace of bitterness in his poetry. For instance, after his release from prison, he was dispatched to Huangzhou, a poor and dreary place, to serve humiliatingly as deputy training officer for the local militia. Yet, on the way there, with such prospects facing him, he still managed to leave us the following gem.

Plum Blossoms* in the Shade

Su Shi

Who holds a cup of wine to cheer what's hidden from the day?
To bloom seems pointless, sadder still to start to fall away.
But luckily, unwilling yet to part, this sparkling brook,
Three hundred turns it makes to keep with me along the way.

* The plum blossom, fragrant, retiring and frost-resistent, was traditionally regarded as a flower of virtue.

A particular appeal of Su Shi is that he seemed to have absorbed and distilled the best of the changes in philosophical outlook brought about by Taoism and Buddhism which had swept through China during the agitation of the preceding period, as is apparent in both the poems *Wildgoose* above and *White Ash* cited next. Though without the emotional intensity of Qu Yuan or the dynamic power of Li Bai, nor yet the tonal subtlety of the one or splendour of the other, poems by Su Shi can give equal satisfaction by often plumbing greater intellectual depths.

White Ash

Su Shi

*All by myself, I'd sound each night the drum and, morning, bell.**
Doors closed, alone, my gaze on fading lamp would nightly dwell.
White ash? - A little stirred, red flames would spring again to life.
Then I, reclined, would hear the rain that 'gainst the window fell.

* Daily routine of a Buddhist monk.

This is not to say, however, that Su Shi was second to anyone in any way when it comes to elegance or ingenuity in "filling in" a *ci*. The following is an example, to which one can find few rivals for sheer beauty.

Where Brows and Eyes Entwine

Su Shi

Where distant hills like puckered brows are gathered,
Like bright eyes some pools of water shine.
And if you ask to which direction the traveller has to go,
It's just where those brows and eyes entwine.
It wasn't long since adieu to Spring we bade.
Must we for your leaving too so soon repine?
But if, once south across the River, you caught up with Spring,
You mustn't then her pleas to stay decline.

The Song dynasty that Su Shi lived in was an age when art and crafts were highly developed. Besides poetry, Song paintings and ceramics must rank among the world's topmost artistic achievements by any standard. It was then also that typographic printing with movable and reusable types was first invented, full four centuries before it made its appearance in Europe. Compared with Tang, however, the art of Song seems somehow to be either constructed on a minor scale or else rendered in a minor key. In place of the gold-green splendour of a Tang landscape, for example, one finds instead exquisite fan-sized cameos of birds or flowers, or else daunting landscapes of sombre colours in which man has but a tiny insignificant role. It was perhaps the difference in fortunes between the two dynasties that is here reflected.

Even at first inception, the Song dynasty never had the vigour of its great predecessor. Let alone expanding Chinese influence as Tang did, it never even managed to recover the great swath of Han territory ceded to Liao before. No institutional foundation was laid to promote or ensure the prosperity of the empire or the well-being of the people. Instead, short-sighted stress was laid only on the continuation of the dynasty. As a result, soon after its inception, the dynasty already found itself in a state typical of a dynasty in decline, with the sophisticated extravagance and corruption of the ruling class supported by limitless and meticulous exploitations of the working population, which were soon followed by repeated cycles of uprisings and repressions.

Only one failed attempt was made at the beginning to recover the lost area ceded to Liao. After that, the dynasty seemed to have lost even the will to resist intrusions, opting instead for appeasement by paying huge indemnities to the invaders. It was not that the country could not defend itself, for however dismal the mismanagment, it still had human and material resources many times beyond what its neighbours could muster. Nor did it lack the will to do so, given that it was a matter of national survival. Indeed, in one major engagement in 1004 A.D., the defenders had already succeeded in putting the enemy in a grave predicament, only to see the appeasers at the top sign away the advantages gained in an ignominous truce by which Song was to pay Liao an exhorbitant annual indemnity. The irony was that the dynasty had always kept at great expense an unusually large standing army which, however, was deployed not to guard the frontiers against invaders but mostly around the capital to guard against internal unrest. Commanders entrusted with the defence of the country were denied adequate resources and hampered by cumbersome rules to ensure that they did not become so powerful as to pose a threat to the dynasty. Even the occasional successes they managed to achieve were sometimes blamed by those above as provocations to the enemy leading to later intrusions.

Fan Zhongyan (989–1052 A.D.), a scholar and soldier outstanding in both integrity and ability, and attributed by later scholars with "the heart of a saint and the pen of a literary genius", was the commander guarding the empire's western frontiers when he wrote the *ci* which follows. Though eminently successful with the meagre resources given him, he was recalled to initiate and lead much needed internal reforms but, barely a year afterwards, was discharged and retired.

Watch on a Lone Fortress

Fan Zhongyan

The scene has changed of late since autumn came to borderland,
But I was not aware when migrant geese for home had flown.
While frontier noises echo round the calls of dismal horns,
The whirling mists at dusk
Close in on fortress wedged among rugged hills alone.
Ah what of distant dreams before a cup of clouded wine?
So long as Yanran stays unconquered, my will is not my own.
And what for the nomad's pipe now hovering o'er the frosted earth
This endless sleepless night?
But my white hairs, and conscripts' tears for homes they once have known.

It was left instead to **Wang Anshi** (1021–1086 A.D.) some decades later to make a more sustained and extensive effort. Wang Anshi was one of those rare original thinkers with ideas far ahead of their times as well as the courage to propagate them against the opinion which then prevailed. He was fortunate too in meeting with one of an equally rare breed, namely an emperor, Shenzong, with the foresight and steadfastness of purpose to let him carry them out. Faced with the situation where the empire's resources were continually being depleted by the expense of keeping a large standing army, the huge indemnities needed for buying off invaders, and the mindless extravagance of the ruling class, Shenzong was naturally attracted to Anshi's claims that all would still be well if only the empire's finances were better administered. To us now, perhaps, his suggestions may appear as no more than common sense. For example, one of the first schemes he put into motion was the so-called *Qing Miao* (Green Shoot) decree, which allowed farmers to borrow, when crops had just been planted and peasant poverty was at its lowest, funds from the government to tide them over that difficult period, to be paid back only when the crops were harvested. The loan would help the farmers, and the interest they paid would enrich the state. True, the interest charged was twenty percent, which to our eyes would seem exhorbitant, but it was still far better than borrowing from private lenders who charged normally five times that amount. The scheme was therefore quite popular. True also, as seen already from this one example, that Anshi's first priority was still, as demanded by the ethics of that time, the service to the dynasty, but he was not without genuine concern too for the peasants, as his following poem shows.

Cricket*

Wang Anshi

In autumn you, who gold-embroidered silk adore,
May sing but those, warm-swathed within, your song ignore.
So you to weavers' homes go, urge yet harder work.
*How many, these, can have a length of silk** in store?*

* Also called *cuzhi* in Chinese, meaning "urging to weave", apparently because the sound it makes is similar to that a shuttle makes on a weaving frame.
** Rolls of woven silk, being light and valuable, were often used along with ingots of silver as a currency of high denomination and as a measure of wealth.

Needless to say, such measures as the *Qing Miao* scheme which cut into their profits were abhorred by the private money lenders, who comprised most of the aristocracy and ruling class, and who often reaped thereby not only the crop but also the land of the hapless farmers at harvest time. Their opposition to the new laws were therefore quite vehement. So the emperor was confronted at one stage with the spectacle of his queen, queen mother, and queen grandmother all pleading in tears, egged on by their respective families, for the new laws to be rescinded. Concessions had to be made and compromises, so that the reforms mapped out by Wang Anshi could never be fully implemented. As he said once to the emperor: "Carrying out reforms in this way is like boiling water, adding fuel with one hand while pouring cold water on it with the other; when can one ever bring it to the boil?" One can thus understand his frustration as expressed in the following poem. Hence, though meeting with some initial successes, he was forced out of office after several years. Emperor Shenzong persisted with the new laws, but without Anshi's guiding hand, they became emasculated of the original potency, only serving to tighten the bond on the peasants without extracting a like contribution from the upper class. And when Shenzong himself died, and the reins of government passed into the hands of his queen mother as regent for his infant son, basically all the new laws instituted by Anshi were repealed. Hearing this, Anshi himself died soon afterwards, a disappointed man.

It has to be said, the opposition that Wang Anshi had to content with came not only from those with mere self-interest at heart, which alone he could probably have survived, but also from principled detractors such as Su Shi, known already to the reader, and Sima Guang, a noted scholar and the author of a monumental classic history of China from the beginning to the Song dynasty.

Visions of Home

Wang Anshi

Through willow leaves, cigalas sing from green shade,
While lotus blooms, by setting sun, give pink glow
On every pool of water there by spring lit -
This white-haired one would wish again to know.

The name, *Zizhi Tongjian,* that Sima Guang chose for his history, which means: A Source-Book for Good Governance, gives us a hint of the way his mind worked, which was indeed quite representative of the period, namely that good governance should be based on tradition, and if tradition does not work for us, it is only because our application of it is at fault. To Sima Guang, therefore, Wang Anshi's innovations were almost poisonous, undermining the moral fabric of the nation, and when he became in turn the prime minister in the queen mother's regency, he made sure before his death that every one of Anshi's new laws, even some which had passed the test of practice, was repealed.

Su Shi's attitude was more measured. Lacking Anshi's originality in political thought, he could not follow where Anshi led, and objected where he saw fit, but he found merit at least in some of his innovations, enough to try persuading Sima Guang to keep them. On the personal level, despite all the persecutions he had suffered from the reformist faction, he had never lost his amity and respect for Anshi. This is perhaps best illustrated by the following *heshi* he wrote to "match" the preceding one of Anshi, written when Anshi had already died and Su Shi himself was in favour at court. A *heshi* is written to match another as a token of amity or respect for the author of the original, with the matching poem keeping often the same format, the same rhymes, and a similar sentiment. Some of the best-loved poems of Su Shi, a master of the art, were *heshi*, which so excelled the originals that these are often forgotten, such as the *Wildgoose* above written to match one of his brother's. This one of Anshi is particularly difficult to match, being very unusual in format (with six characters in each line) and halting in rythm, reflecting both the originality and some awkwardness in his personality, but Su Shi managed all that with flying colours.

One Autumn Morning

Su Shi

O' er stream and meadow, autumn-tinged a clean shade,
By rain washed, morning breeze and sun spread fresh glow.
From hence, retired back home, I would, to deep south.
Who's left, to see me off - who did me know?

With reforms petered out but not the rivalry between the two political factions, the Song dynasty probably would not have survived had not its main tormentor, Liao, soon suffered also a decline. But when Liao was supplanted by a vigorous new dynasty, Jin (to be distinguished from Jin, the Han ethnic dynasty some centuries earlier) of another ethnic group in 1125 A.D., Song itself came immediately under threat. The stiff resistance put up by some generals and the population at large was repeatedly undermined by the appeasers at the top who still dreamt that the invaders could somehow be bought off. By 1127 A.D., however, the Song capital fell and the reigning emperor together with his father and predecessor (who, the year before, had hastily abdicated and fled to the south for safety when danger first threatened, returning only after the danger was thought to have receded) were taken captives and transported north to pass the rest of their lives as prisoners.

The emperors got what they deserved, but what about the people remaining now under the invaders' iron fist? The following two poems give a glimpse of what they had to suffer.

Jiang Xingzu was a county chief around 1126 A.D. in what is now Henan province. He was killed in defence of the county against the Jin invaders and his **daughter** was captured and carried away as a prize. On her way north, she left this poem on the wall of one of the overnight stops at Xiongzhou. At that time, she was just over fifteen years of age, and was noted for her beauty and her poetic talent. And this is apparently all that is known about her. Yanshan far to the north was then the centre of Jin power. Were the poetess from an illustrious family, she might have been ransomed, but given that she was only the daughter of a county chief who was in any case already dead, this was unlikely.

Found on a Wall of Xiongzhou Station

Jiang Xingzu's daughter

Cross country drifts the morning cloud,
While like a flowing stream unceasing, groan the cartwheels loud.
Pale grasses grow in yellow sand;
A cold moon shines on village where but two, three houses stand.
With wild-geese screaming past in flight,
A thousand pains rake piercing through my heart both day and night.
As nearer Yanshan now we turn,
The more for home my heart will yearn where I can ne'er return.

Li Qingzhao (1081 A.D. – ?), in contrast, came from a distinguished family. Propably uncontested as the greatest poetess that China has ever known and having but few rivals, male or female, in the *ci* genre that she mostly wrote, she was recognized in her lifetime already from an early age. Married at eighteen to Zhao Mingcheng, a noted scholar of antiquities with also a successful career in the imperial service, who shared her passion for learning and, though not her equal in poetic talent, was good enough at least to appreciate hers, she had a happy early life. The conjugal regard between the two extended over both the emotional and intellectual spheres, which was rare in those days when few women were highly educated. The story goes that in their happy days together, they would often spend an evening challenging each other on quotations from ancient texts as a game, the prize being the first serving from a pot of fresh-made tea. Once, when Mincheng was away on official duty, Qingzhao sent him a *ci*, and he wanted so much to write one worthy of hers in reply that he shut himself up in his study for three whole days and nights, declining all visitors and duties. As a result he produced some fifty pieces, which he showed to a discerning friend with Qingzhao's mixed in for the friend to pick the best. The choice fell unequivocally on two lines of hers

When the Northern Song capital fell, Mingcheng fled with his family to the relative safety of the south, but he died soon afterwards. Qingzhao lived on for some more years in exile when the following poem was written, but it is not known exactly where and when she died.

The Festival of Lights

Li Qingzhao

The setting sun is molten gold,
The evening cloud, compacted jade,
But where is he?
Dense mists have tinged afresh the willows;
The plaintive flute appeals to budding plums.
How many signs of spring there be?
Again, the Festival of Lights!
The weather is, as usual, mild,
But will there not be wind and rain to some degree?
These carriages fine,
All empty shall return,
That my poet-friends have so kindly sent for me.
It used to be, on this Fifteenth Night
Of the First Month and of the Year,
In happier days when both the country and my heart were free,
All decked-out in finery,
To rival others I would go,
Fingering, green-bonneted, a golden tassel like a willow tree.
But now I'm old, with hair
Like wind-swept mists unkempt;
Evening outings I no longer wish to see.
Just sit behind the curtains and hear
Some snatches of the laughing words of others,
Would more with my present mood agree.

Meanwhile, after the fall of Song in the north, whatever remained of the rotten hierarchy fled south and established as emperor a brother of the one taken prisoner. Thanks to the fierce resistance to occupation put up by the people in the north, this Southern Song dynasty was given a little time to catch its breath. After some initial scares which saw the emperor escaping further south by sea, they settled eventually on the beautiful lakeside city of Hangzhou as capital.

They called their new capital Lin'an, meaning temporary comfort, which was apparently all they sought. Thus although at one stage, when the Jin dynasty was itself in disarray because of internal dissentions, a real chance presented itself for the Song forces under some able commanders of regaining the lost land in the north, their efforts were deliberately scuppered by the appeasers around the emperor who was himself afraid of losing his throne if the Song forces proved too successful, leading to the release of his predecessor who was still living in captivity. The advancing forces were recalled, and a shameful treaty concluded, by which Song was to cede to Jin more territory, abandoning thus even more Han people to foreign rule, and to pay Jin an even bigger annual tribute, which expense had of course to be borne in the end by the people.

Yue Fei (1102–1141 A.D.) was the most successful of the generals leading the advance to the north around whose banner rallied the resistance forces struggling for release from captivity. Since early youth, he had set his heart on regaining the lost land for the nation, as evidenced in the following *ci,* and indeed in almost everything else he wrote whether as a poet or as the noted calligrapher that he was, and it did look at one stage that he was about to achieve his goal.

On Brown Heron Tower*

Yue Fei

In the distance through the haze, I see the many-citied
Central Plain before me spread,
And recall the former days
When willows and flowering trees their shade
On grand Imperial Palace shed.
Sweet music from the court is wafted through the balmy air;
Prosperity and peace are showered on each passing head.
But now, mailed horsemen roam and trample over all the countryside,
And fill the air with dread.
Our people, they
In fosses tread.
Our soldiers, where?
To spear-points fed.
The lay of land looks still the same, but hides
A thousand villages all ruined, dead.
Ah when allowed to lead a valiant troop across, in one
Clean sweep, clear all this filth from off the Yellow River bed?
Then come again, resume my present leisure tour - on heron's
Back to seek oblivion fled.

* (in present day Wuhan) where legend has it that a sage of old attained immortality and flew away on the back of a brown heron.

But then came the general recall, which would have left Yue Fei with his flanks exposed even if he could overcome his scruples of absolute obedience to the emperor which ethics then demanded. Reluctantly then, he returned to the capital, where he was relieved of his command, put in prison on some trumped up charges of treason and covertly killed together with his son and a lieutenant. His death was much lamented by the nation and even today, one still finds in Hangzhou a shrine in his honour, at the entrance to which are found a couple of kneeling figures representing his ignominous accusers.

With Yue Fei and others like him suppressed, the Song dynasty lost all credibility as representative of the nation, and held on to its rule of the south only as a vassal of a foreign lord. The mood of those remaining who still had the nation at heart was one of despondence, as summed up by the following *ci* of **Xin Qiji** (1140–1207 A.D.). Qiji was himself in his youth a freedom fighter resisting the Jin occupation in the north, where he was known for some very daring exploits. With less than no support from the dynasty in the south, however, the resistance petered out and he was forced to retreat with his remaining forces to the south. There he seemed to have been made much of and honoured for some time, with a position at court but no power, and eventually grew old leaving us little more than his poems

Indeed, with no pretence at self-defence, the Southern Song dynasty would not have survived had the Jin dynasty itself not followed the same pattern of decay and found itself soon in a similar state. And so the two dynasties in the north and south coexisted and dragged on for some more decades until the Mongol explosion in the north which shook the ancient world as far as Eastern Europe broke south also and swept away both.

O Sorrow, Why Dost Borrow

Xin Qiji

In youth, I did not know how Sorrow tastes.
Then up tiered towers I loved to go,
Then up tiered towers I loved to go.
To write new poems mourned I Sorrow so!

But now, full well I know how Sorrow tastes.
Say Sorrow yet again, or no?
Say Sorrow yet again? Or no,
Just say: cool air befits the autumn glow.

Roundoff

We have now come to the end of our story. The history of China goes on, of course, but the development of classical poetry, which we have made the vehicle for the present narrative, essentially stopped soon after the fall of the Northern Song dynasty. It was not that in the centuries that followed there was any lack of fine poets or of good poetry, for poetry, no doubt, is a necessity for any age, but that the prolific pace at which new poetic forms were generated through the twenty centuries since its first inception in the Spring and Autumn period appeared now to have lost its momentum. The poets of the subsequent dynasties seemed to find the existing forms already adequate for all they wanted to express, and saw no need to develop any more of their own.

This sudden arrest in development of poetry at the end of Northen Song cannot easily be ascribed to the domination by "foreign", that is, non-Han, powers, as China often was since then, first by Jin and then by the Mongols for two, and later again by the Manchus for another two and a half centuries. China has seen such set-backs before, notably in the three or four hundred years after the breakup of the Han dynasty, or the century or so after the demise of Tang. The invaders were eventually simply amalgamated into a grander, more varied Chinese culture, still mostly Han in character because of the sheer weight of the Han population, but now greatly enriched by the influx of new values and new ideas. Witness the blossoming during the Tang dynasty, with the "foreigners" themselves soon producing fine poetry in the Han tradition. One could perhaps even say that the development of new poetic forms had been, at least in part, driven by the "foreign" invasions.

This time round, however, the process was not repeated. In the Yuan dynasty under the Mongols, a new genre called *qu* did indeed flourish for a while, derived from the Song *ci* following the same general principles, only with the templates *cipai* replaced by the *qupai.* Composed still mostly by the educated class, but now deposed from their erstwhile elevated position, the Yuan *qu* possesses a refreshing earthiness both in language and in setting, but cannot compete either in sophistication with the Tang *shi* and Song *ci,* nor yet in freshness and spontaneity with such genuine products of the common people as, for example, the *Shijing* or the *yuefu* of the early Han dynasty. Thus, though not without their own admirers among the discerning, the Yuan *qu* has never quite acquired the same significance as poetry in the popular imagination. And when the country returned to Han control under the Ming dynasty (1368–1644 A.D.), despite the amalgamation of culture which had since occurred, poetry proper reverted back to the earlier forms, with the Yuan *qu* siphoned off to develop instead as the Yuan and Ming plays, a discipline with a different appeal and norm. Poets of Ming and subsequent dynasties were still content with the forms developed in Song and before.

Perhaps the reason is that although society in that period was changing fast in many ways, the formal structure was not largely altered. At the top of that society was the Emperor, whether Han, Mongol or Manchu, and everything else followed from there. Indeed, even the conception of Heaven (on which speculation the Chinese, singular perhaps among ancient peoples, spent very little of their time) ran roughly along the same lines, namely just a celestial version of the imperial system on earth. We moderns, imbued with current ideas of democracy and individual rights, tend to criticize the ancients for servility in submitting to such a system. This may be unfair, for surely, they lacked no courage, the best among them, no more than we do, to stand up for what they believed, and history is full of such examples. They submitted to the imperial model of society only in the belief that it was part of the natural order, on which the stabilty of society and even of the world depended. And that model had served China well in a sense; witness the stability that China had achieved relative to other nations, over a period of some four thousand years. History was thus viewed as largely static, where later generations just repeated what had gone before, sometimes better, sometimes worse. That being the case, what sufficed for the past would also suffice for the present. It was so for government, and should be so also for poetry.

At least, most seemed to think so, but if they did, then they were wrong. Something appeared to have happened in the Song dynasty which was fundamentally changing the society. Despite the dynasty's own weakness and inept governance, society under Song did not stagnate, but seemed rather to have acquired a life of its own and advanced rapidly on many fronts, independent or even in spite of the management from above. Besides arts and crafts, trade flourished, and farming techniques improved, supporting an ever increasing population. Indeed, apart from relatively brief interruptions occasioned by the changes in dynasties, the increase in population has been maintained steadily since, through all the ensuing centuries, until it reaches the staggering 1.3 billion of today. It would almost seem that by the Song dynasty, Chinese society had grown to adolescence, shaking off at last the "parental guidance" of the Emperor, which was in any case found by that time to be woefully inadequate.

Adolescence, however, is a difficult and busy time with a lot of learning and growing to do. And China's adolescence was a particularly difficult one for several reasons. First, China is and always has been poor in natural resources in relation to the population it has to support. As of today, China has per capita only one-third of the world-average in arable land, a quarter in renewable water resources, and one-eighth in forest area at its disposal. This poverty was in evidence already by the middle of the Han dynasty and, throughout the centuries since then, it had always required a lot of hard work, ingenuity, self-discipline and even self-sacrifice from all to feed and clothe the whole population, keeping it in moderate comfort and at relative peace. Indeed, it was an achievement, supporting so large a population on such meagre resources, which had perhaps never been paralelled anywhere else at any time in the history of humankind. Secondly, as already mentioned, China since the Song dynasty had to manage the amalgamation of cultures occasioned by the influx of the Mongols and the Manchus. The difficulty seemed particularly acute with the Manchus who founded the Qing dynasty, for they sought not just military and political domination as earlier conquerors of Han China did, but even the imposition of their own culture above that of the Hans, leading to an odd artificial mixture which can at times be singularly depressing. But thirdly, perhaps worst of all, when the Qing dynasty, following the pattern of its predecessors, had declined to its terminal stage and become ripe for replacement, the situation was complicated by the great happenings then in the outside world. In particular, the newly industrialized western nations, taking advantage of the Qing dynasty's decadence, were encroaching more and more on Chinese sovereignty. Britain went as far as sending gunboats to force on China the infamous opium trade. In 1900, a coalition of eight western nations invaded China and sacked the imperial palace in Beijing, when only the rivalry among them saved China from following the fate of India in sinking to the status of a colony.

The Qing dynasty was finally deposed in 1911 and a republic was declared, but it took some time before a semblance of order was established and effective control of the regions wrested from the warlords. Soon after it was more or less again united, however, the Japanese invaded, as part of their militarist dream for a Great East Asian Empire, and occupied most of the populous eastern half. This came to an end in 1945 at the conclusion of the Second World War with China on the winning side, and Chinese sovereignty was restored. But then civil war broke out between the governing Nationalists and the insurgent Communists, only ending with the foundation in 1949 by the latter of the People's Republic. Even after that, although living conditions for the people soon vastly improved, China had yet to go through several major upheavals, such as the failed Great Leap Foward followed by famine, the chaotic Cultural Revolution, and the Tiananmen events which rocked the world as recently as 1989, before settling down to the relative quiet and prosperity we see today. Unfortunately, with some basic problems unresolved, it is still unclear that the Chinese nation will henceforth progress and develop on an even keel. But now, at least, there is hope.

Hope for its own future alone, however, is no longer enough, for when China awoke from its trauma of adolescence, it found itself in a different world. It used to be that, esconced in its bountiful homeland of two great river valleys surrounded by the desert and the sea, the forces affecting its life were mostly centripedal. True, it had often to defend itself against "invaders" and adapt itself to "foreign" norms, but in time it realised that what it had thought were foreign were but previously unrecognized parts of itself. But this time, it is different. It finds itslf now as a member of a community, where there are others completely distinct from itself, whose company it has learned to value, which it can in any case no longer avoid. In other words, adulthood has opened its eyes to a new world of a promised richness previously unimagined and unimaginable. But, as usual, adulthood also brings with it new responsibilities. It has now to ask itself, and will be asked by others too, what is there that it could contribute towards the common good. Like an outsized adolescent newly out of school, China emerging now as a world power could not help but feel some embarassment in seeking its right place in society, and in clarifying for itself what the rest might expect from it.

It is often said that some hardship suffered in adolescence may contribute towards a responsible, happy adulthood. So may it, one hopes, for China, as it now enters the world society with two hard-won offerings in its hands. One is *frugality*, which it has learned enforcedly from poverty; the other *tolerance* (we are speaking here of the culture, not of the present or any particular government), which it has acquired partly as an inheritance and partly by necessity to survive adversity.

"Waste no water even when you live by the river," Chinese children are traditionally taught. It is this dictum of frugality which has helped the nation survive through the centuries despite meagre resources. Even now, China as a whole has a higher savings ratio than most other countries, which some say will help it ride out better the current financial storms. Though born of human necessity, this frugality has in time acquired also an extra-human dimension in the form of a general regard or respect for nature; witness the predominance of landscape paintings and sketches of birds, flowers and other bits of nature in Chinese art. And although the Chinese culture shares with most others the delusion that humanity is at the apex of creation, it has not succumbed to the notion of some that the rest of nature exists only for our exploitation and our use. In other words, nature is there not just to be our environment, which all agree we have to preserve for our own good, but is an entity in itself from which we draw our sustenance and of which we are but a part. Thus, with the injunction on children not to waste water comes no explanation that the water may be needed by some other humans. Rather, the implication seems to be that water is a bounty granted to us by nature, and as such already is never to be abused.

The inscription on one of the seals left by the author's father reproduced on the dedication page, the second from the left, says in part: "Repudiate not others just because they differ from you." This attitude of tolerance, so different from that epitomized by the assertion: "If you are not with us, you are against us," which some propagate, causing such havoc in our present world, was partly inherited by the Chinese nation from its history as a child of many cultures. But it was also partly imposed on it by the force of circumstances, for how else could the Han nation have survived the century-long domination by the Mongols, say, with so different social and ethical norms? It is this tolerance which allowed such erstwhile "foreign" elements as the Di and the Xiongnu, and many other ethnic groups before and after those, to merge so seamlessly into the Middle Kingdom culture that only experts in ethnological research, and most often not even they, can tell them apart or, as they say, whether a Mr. Liu one meets is a Liu as in Liu Bang, the founder of the great Han dynasty, or a Liu as in Liu Yuan, a ruler in the Northern Dynasties of Xiongnu descent. It is also this spirit of tolerance which could explain, at least in part, why the influx of the great religions from the west had caused in China only ripples of discord compared with the huge upheavals they gave rise to elsewhere.

And these attitudes, frugality and tolerance, are they not exactly what humanity may need today to combat the two most serious threats to its continued existence, namely, the rapid depletion of natural resources leading to a degradation of the environment, and the distrust between cultures, threatening to errupt at any time into violence as terrorism and war?

Returning now to poetry, we see that with a long series of humiliations, deprivations, and difficulties to contend with, it is perhaps not surprising that adolescent China had not found the voice to sing with the same innovative eloquence as it did in its infancy in the centuries before. Indeed, it was not until the twentieth century when China came into close contact with the rest of the world, forcing a rapid change on society and on the ideas that sustained it, that the need was felt and effort made to develop new poetic forms. One can discern two trends. On the one hand, some would feel that in order to express the new ideas and the emotions they generate, one would need a radically new poetic form also. This feeling no doubt reflects the nations's awakening to foreign ideas, and to its own failings of the last century which sapped the confidence of many in all things traditional. Thus some would discard altogether rhymes and tonal structure which formed the backbone of classical Chinese poetry and embark on a version of the western blank verse. On the other hand, some would maintain that the classical forms, though overly restrictive in certain circumstances, contain enough variations among them to allow adaptations for modern use and that they have already the capability to express whatever new ideas and emotions that the modern poet may choose to express.

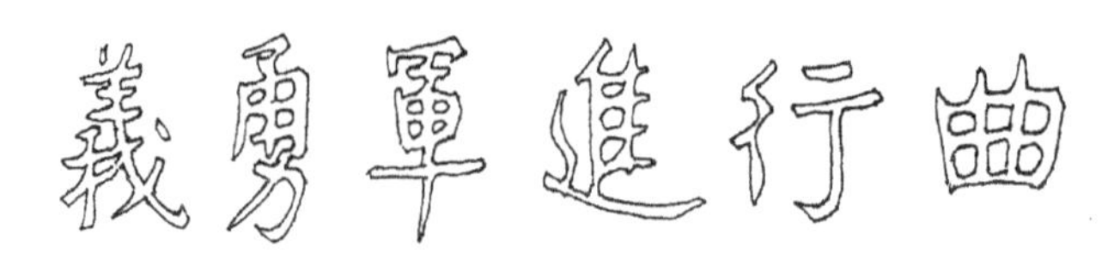
義勇軍進行曲

中华人民共和国国歌

田漢 詞 聶耳 曲
Stand tall! We'll be no sla - ves, we all.

4
With our flesh and blood we'll build us our mod-ern Great Wall.

9
Th'Chi-nese nat-tion's fac -ing a most dan-ger - ous time.

14
Till our last breath we will an-swer na-tion's ur-gent call. Stand

The March of the Volunteers
(Now the National Anthem)

Tian Han (words)   Nie Er (music)

This Festering Wound of History
(Song heard at a London rally, June 3, 1991)

You think we see not just because our eyes are shaded?
Or that we'll never hear if you but stop our ears?
The Truth we know because our hearts they feel the anguish -
O, how much longer still have we to bear in silence
This festering wound of History? Ah ....!

If there are tears, they'll serve to wash away the dust stains.
If our blood be warm, it'll be for freedom a worthy price.
We'll let tomorrow know what we today have suffered.
Our cries of anger will expose to all the world
This festering wound of History. Ah ....!

Dark Tears

Yau Shun-chiu

So what is left, but the prints of exiled feet,
And echoes of the hoarse protesting voice, -
And some in fetters for no crimes detected
Leaving wives and daughters unprotected, -
And blood so warm that it has never dried
Though spilt long since from hearts whose pulse was stopped?
And what for me o'erseas in distant lands,
But ink-stains one by one on paper dropped?

Ambition of the Common Man

Lunxizhishi

But shoulder we our trust -
The nation's rise or fall.
Let not fair fame, we common folk,
Our hearts enthral,

For heroes are but dust,
To dust they all return.
What boast today or hope tomorrow,
Encased in an urn?

On the preceding pages are a few random examples of relatively recent poetry, selected on no better criterion than that the author had, for some reason or another, translated them. Though far from representative, they nevertheless reflect already China's painful recent history. The first, *March of the Volunteers,* now the national anthem of the People's Republic, was written during the resistance to Japanese invasion, and the others were connected to the Tiananmen events of 1989. Of these, the first three are "nontradionalist" and the last is "tradionalist", but in English rendering, of course, the difference is not easily discernable.

As to which approach, the "traditionalist" or the "nontraditionalist", is the correct one, only time will tell. For the present, I think it is fair to say that neither of the above approaches have so far attained its goal in that neither has yet been able to supplant the Tang poets, say, as poetry in the popular imagination. When something occurs to trigger a poetic sentiment, it is still a line from a classical poet that is recalled, not one from a modern poet of either school. For instance, when the present Chinese premier Wen Jiabao, who won widespread appreciation for his behaviour and action in relief efforts for the devastating Sichuan earthquake of 2008, was interviewed soon afterwards at a meeting of the United Nations, he chose, quite movingly, two lines from the Tang poet Li Shangyin to express his feelings and sense of dedication. These were the two lines rendered as:

Till death, the silkworm winds itself in endless bonds of silk,
And only when to ashes burnt, a candle's tears will dry.

from the poem *Untitled* cited above.

For the "nontraditonalist" poem, a problem is that, without either rhyme or tonal structure, it is extremely hard to memorize. For myself, I can say that all the poems I have here included were translated from memory, which is, I find, the most effective way for reproducing the feel of the original. The translations were carried out over a period of perhaps twenty years, with some dating even earlier. And nearly all have remained in my memory since, word for word, except for the couple of "nontraditionalist" examples here included, of which I have been able to retain only the general feel. For the lover of classical Chinese poetry, one great pleasure is to be able to recite aloud a piece from memory when moved to do so, just as one would sing a song to oneself when one feels the urge, and this pleasure is denied, to myself at least, with the "nontraditionalist" poem, whatever its other qualities.

For the "traditionalist" poem of a modern poet, on the other hand, the problem is that, however new the ideas it explores and the sentiments it imparts, one still feels, both in the writing and the reading of it, that one is missing something in freshness and originality.

Most likely, the truth lies somewhere in between, namely that new forms need to be developed which loosen the often stifling conventions that centuries of tradition have imposed on the ancient forms, but which still retain such seemingly intrinsic and immutable values of classical poetry as tonal structure, bound to the special characteristics of the language.

With China now approaching maturity, and entering perhaps a new golden age, we can hope that new poetic forms too will soon emerge to satisfy in full our modern needs.

Appendix I: Timeline and Maps

Intended merely to help the reader place the poems and events cited, the sketched timeline and maps here cover only the period referred to in the text and can lay no claim either to completeness or accuracy. Due to lack of space, the events listed in the timeline are necessarily limited, the selection being often decided only by personal taste. The maps, though based on the authoratative collection published in 1982 by the Chinese Academy of Social Sciences edited by Tan Qi Xiang et al., are but very imperfect copies of the originals, they being so severely constrained in size.

An outline of Chinese prehistory is given in the text. There were two historical dynasties before the Zhou, namely the Xia (21st to 16th century B.C.) and the Shang (16th to 11th century B.C.). Thus, by the time the Xia dynasty began, the Pyramid of Cheops had already been standing in Egypt for some five hundred years, which places the Chinese civilization much behind the Egyptian. The Shang dynasty was a rough contemporary of Ramses II (the Great) in Egypt, with the Chinese civilization by then beginning to catch up in sophistication with its Egyptian counterpart. But whereas the Egyptians built in stone, the Chinese built in wood and mud leaving thus few spectacular monuments. The Shang bronzes, however, are in a class of their own and bear witness to the level of culture then achieved. After that came the Zhou dynasty, lasting for nearly eight hundred years, in the second half of which our narrative here begins.

Year	Period	Events in China	Events Elsewhere
800 B.C.	**The Spring and Autumn Period**	***The Seventh Month*** ^ Zhou capital moved east to Luoyang Iron tools appeared	*Homer* ^ The first Olympics Sargon II of Assyria at Nineveh
700 B.C.		Expansion of Han ethnic influence *First record of Halley's comet*	Library at Nineveh with *Epic of Gilgamesh* Zarathustra in Persia Solon at Athens
600 B.C.		***Shijing*** *(poetry from 11th to 6th century B.C.)* Laozi (Lao Tze) Kongzi (Confucius)	Nebuchadnezzar the Great at Babylon *Thales "predicted solar eclipse"* The Buddha in India Cyrus I of Persia *Persepolis*
500 B.C.	**The Warring States Period**	***Zuozhuan*** *(History of China of the period)* Consolidation, rapid expansion of "Middle Kingdom" culture	Battle of Marathon Battle of Salamis Pericles at Athens *The Parthenon* *Herodotus* *Sophocles, Euripedes* Socrates, Hippocrates
400 B.C.		"Flourishing of a hundred schools". (academy at Linzi with "10000 scholars") Blast furnaces in use for high quality steel ***Qu Yuan***	Peloponnesian War *Thucidides* Plato founded academy *The Bhagavatgita* *Praxiteles* Alexander of Macedon

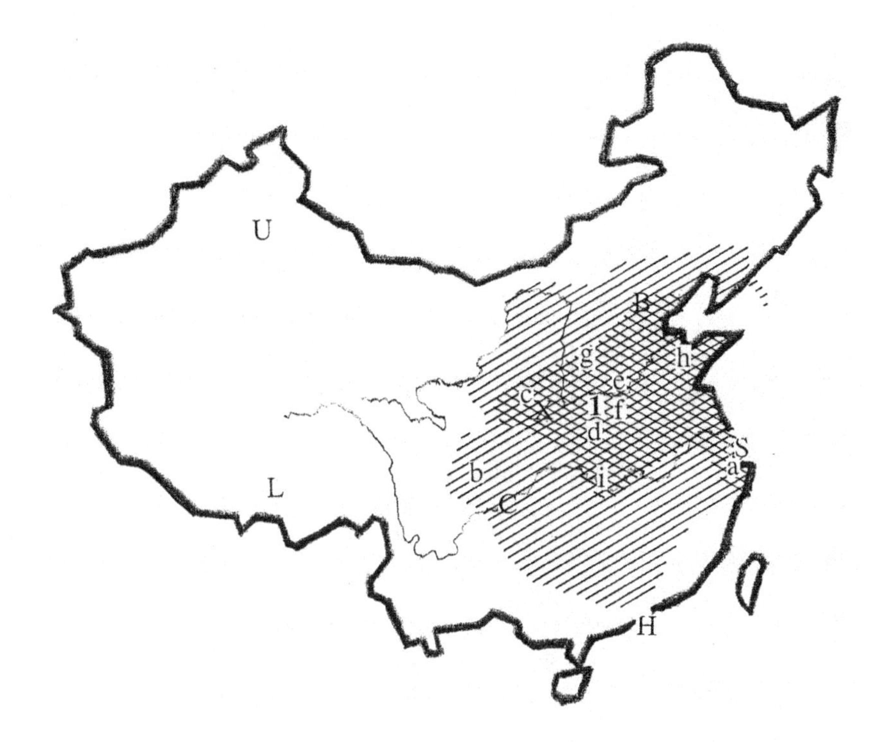

Modern cities for reference: B = Beijing, C = Chongqing, H= HongKong, L = Lhasa, S = Shanghai, U = Urumqi, X = Xi'an.

▧	Rough area spanned by feudal states in the Spring and Autumn period..
▨	Approximate area occupied by the Warring States in the 3rd century B.C..
1	Luoyang, capital of the Eastern Zhou kingdom.
a	Liangzu culture, 3000–2000 B.C.
b	Sanxingdui, early Shu culture, c. 2000–1000 B.C.
c	Scene of *The Seventh Month.*
d	Scene of *Rouse not the Dog.*
e	Scene of *War-Drum, Fluff-Ball, Winning Smiles.*
f	Scene of *Tuck up Your Clothes.*
g	Scene of *Tightly Bound.*
h	Linzi, capital of Qi, with about half-a-million inhabitants and academy with "ten thousand scholars".
i	Ying, capital of Chu, scene of Qu Yuan's *Lament for the Fall of Ying.*

Year	Period	Events in China	Events Elsewhere
300 B.C.	**The Warring States**		*Euclid in Egypt* *Archimedes in Sicily* First Punic War Asoka in India *(edicts)*
	Qin	Shihuang unified China *The Great Wall* *The Terracotta Army*	Hannibal in Europe The Ptolemys in Egypt *Library of Alexandria*
200 B.C.	**The**	Exemplary reigns of Wendi and Jingdi Zhang Qian's first mission to west Great expansion under Emperor Wudi	Expansion of Rome
100 B.C.		*Shiji, comprehensive history of China by Sima Qian* *First record of sunspots*	Julius Caesar Augustus, Emperor of Rome
0	**Han**	Pingdi census gave 13 million tax-paying households	Jesus Christ Claudius census gave 7 million Roman citizens
100 A.D.	**Dynasty**	Ban Chao opened way to west, sent envoy to Rome who got as far as the Levant (?) *Invention, manufacture of paper by Cai Lun* ***Cao Cao***	Trajan, Roman Empire at maximum extension Exemplary reigns of the Antonines at Rome

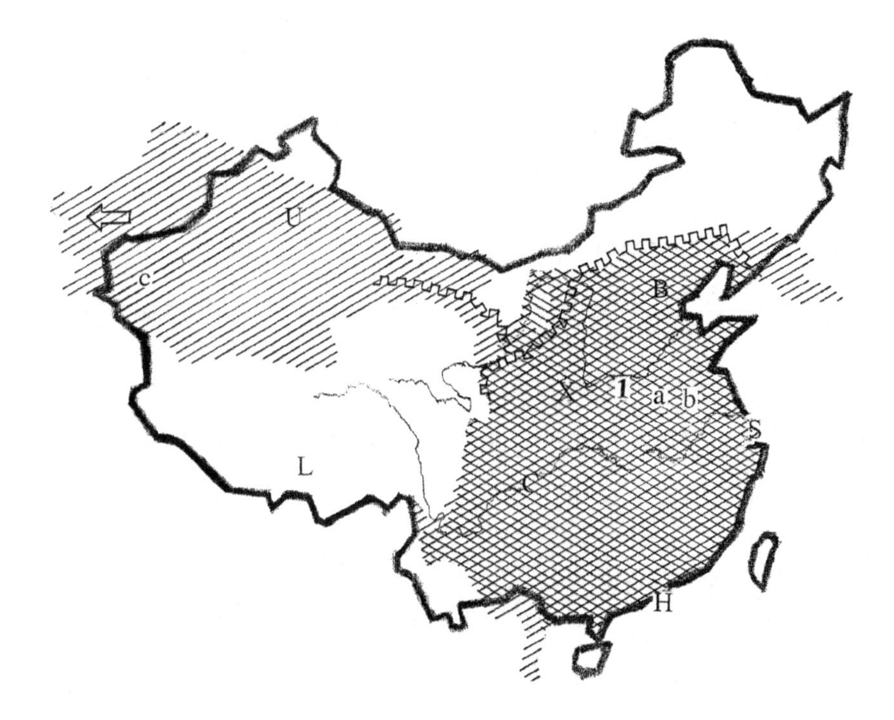

Modern cities for reference: B = Beijing, C = Chongqing, H = HongKong, L = Lhasa, S = Shanghai, U = Urumqi, X = Xi'an.

- The Qin empire.
- The Han empire at its peak in the first half of the 1st century B.C.
- The Qin Great Wall plus Han extension westwards outside Qin boundary.

X — Xianyang, capital of Qin, Chang'an, capital of Western Han, and the Terracotta Army were all within the area covered by the letter X on map.

1 — Luoyang, capital of Eastern Han.

a — Chen Sheng's uprising started here.

b — Scene of Xiang Yu's last stand.

- Zhang Qian's historical mission to Yuechi by River Amu. Captured at start and kept prisoner by Xiongnu for ten years, he escaped to complete mission and bring back first information of nations as far west as Rome.

c — Han garrison at Kashgar where Ban Chao mostly operated. Mainly by wise diplomacy, projecting rather then applying Han power, he gained the adherence of neighbouring nations to secure passage to the west.

Year	Period	Events in China	Events Elsewhere
200 A.D.	**Han**	Battle of Chibi kept country divided	Sassanid dynasty began in Persia
	The Three Kingdoms		
	West Jin Dynasty	Envoy received from Rome	Diocletian split Roman Empire
300 A.D.			
	East Jin Dynasty	Northern China overrun by invaders	The Edict of Milan Constantinople founded The Gupta dynasty began in India
		Grottos, paintings of Dunhuang begun	
400 A.D.		Battle of Feishui kept Han south independent	Fa Xian in India
	The Northern and Southern Dynasties	***Tao Yuanming***	The sack of Rome by Alaric, the Goth The Huns in Western Europe The Vandals ravaged Europe and Africa
		Spread of Buddhism	
500 A.D.			The Anno Domini system introduced
	Sui Dynasty	Examinations for recruiting officials first instituted	
600 A.D.	**The Tang Dynasty**	The exemplary reign of Emperor Taizong	Mohammed at Medina Moslem expansion
		First recorded advent of Christianity	Xuan Zhuang in India
		First advent of Islam Empress Wu Zetian	Omayyads at Damascus

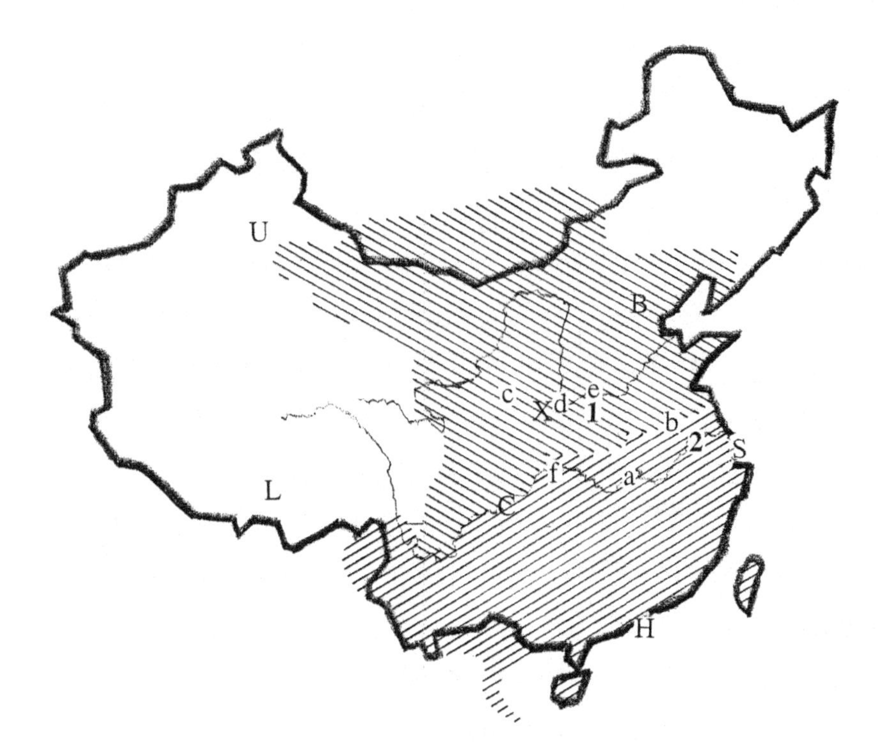

Modern cities for reference: B = Beijing, C = Chongqing, H = HongKong, L = Lhasa, S = Shanghai, U = Urumqi, X = Xi'an.

Qianqin, the northern contestant, under Fu Jian of an ethnic minority, and

Eastern Jin, the defender of Han culture, at the Battle of Feishui, 383 A.D.

1 Luoyang, capital of Eastern Han.

2 Jiankang, also called Jianye and Jinling, capital of Eastern Jin and most southern dynasties.

X Chang'an, capital of Tang, was about 20 km south of the modern city of Xi'an and therefore within the area covered by the letter X on the map.

a Battle of Chibi, defeat of Cao Cao led to separation into Three Kingdoms.

b Battle of Feishui, Eastern Jin won and kept Han south independent.

c Longtou, scene of *Overnight Stop at Longtou.*

d Mount Huayin, scene of *The Well on Top of Mount Huayin.*

e Scene of *Dancing by the River.*

f Sanxia, the Three Gorges, scene of *Ballad of the Three Gorges* and of Li Bai's *Down the River to Jiangling.*

Year	Period	Events in China	Events Elsewhere
700 A.D.	**The Tang Dynasty**	Exemplary reign of Emperor Xuanzong ***Li Bai, Du Fu*** Capital Chang'an overrun by rebels	*Beowulf* Tours, Charles Martel halted Arabs in Europe Abbassids at Baghdad
		Chang'an occupied by Tibetan invaders	Mayan "First Empire" in Central America
800 A.D.		***Bai Juyi***	Charlemagne, Holy Roman Emperor
		Du Mu, Li Shangyin	Mayas at Yucatan
		First printed book	Romance languages evolving from Latin
900 A.D.	**The Five Dynasties**	Mass printed books by state agency ***Li Yu (Li Houzhu)*** First rockets used in warfare	Maximum extension of Byzantine Empire
	The Northern Song Dynasty	First banknotes used in Sichuan	
1000 A.D.		First record of a supernova First printing with movable types Wang Anshi reforms *Zizhi Tongjian by Sima Guang* ***Su Shi (Su Dongpo)***	Viking discovery of America *The Vinland Saga* Norman Conquest of Britain *Bayeux tapestry* *University of Bologna* First Crusade
1100 A.D.		***Li Qingzhao***	
	The Southern Song Dynasty		Second Crusade *Chanson de Roland*
		Xin Qiji	*Omar Khayyam* Genghis Khan

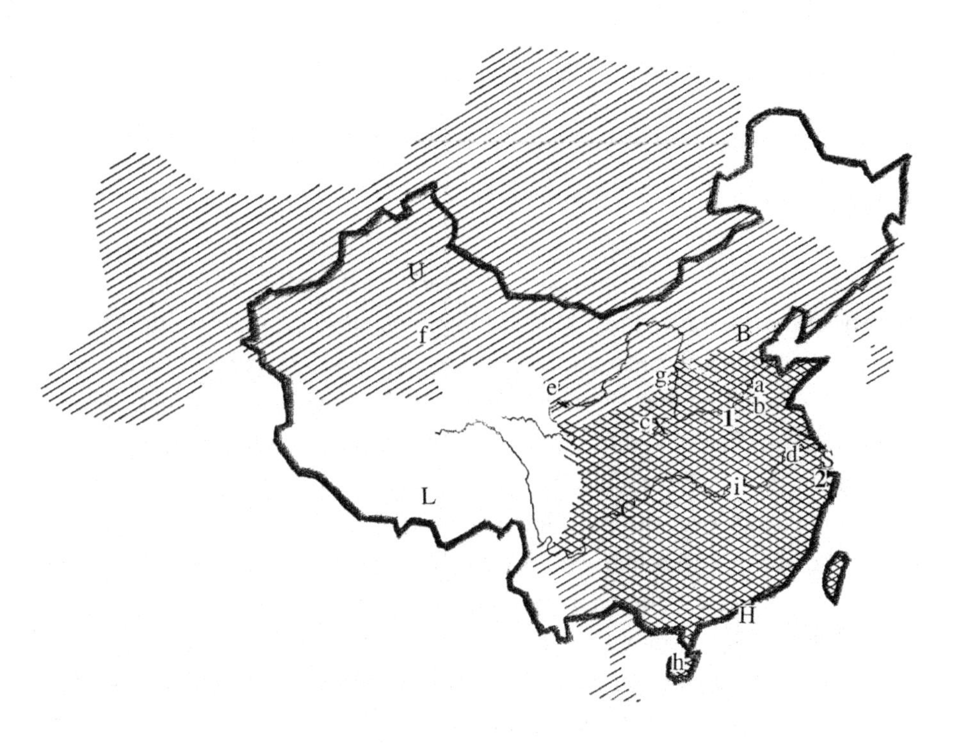

Modern cities for reference: B = Beijing, C = Chongqing, H = HongKong, L = Lhasa, S = Shanghai, U = Urumqi, X = Xi'an.

(diagonal hatching)	The Tang Empire at about its maximum extent in 669 A.D.
(diagonal hatching)	The area under the Northern Song dynasty.
1	Kaifeng, capital of Norhern Song.
2	Lin'an, capital of Southern Song, scene of *Shoo away the Nightingale.*
a	Qufu, scene of Li Longji's *Worship at the Shrine of Master Kong.*
b	Taishan, scene of Du Fu's *Taishan Peak.*
c	Scene of Wang Wei's *Pastorale.*
d	Changgan, scene of Cui Hoa's *Chance Encouter Afloat.*
e	Qinghai, battleground between Tang and Tibet mentioned in Du Fu's *Chariots Rumble.*
f	Milan, ruins where Kan Man Er's *Wolves of the Manor* was found
g	Scene of Fan Zhongyan's *Watch on a Lone Fortress.*
h	Hainan Island, Su Shi's last place of exile.
i	Wuhan, scene of Yue Fei's *On Brown Heron Tower.*

Our narrative in the main text ends soon after the fall of the Northern Song dynasty in the twelfth century A.D. At that time, as one can see in the timeline above, China was a fair amount ahead in many things compared with a Europe still recovering from the "barbarian invasions" of the preceding period. For instance, paper, first manufactured by Cai Lun in early second century A.D. and in common use in China since, was introduced to the West by the Arabs only after their victory over the Tang imperial army in 751 A.D. in a battle near the present border of Kazakstan with Kyrgyzstan, in which some Chinese artisans were captured. And typographic printing with movable and reusable types, first invented by Bi Sheng in around 1041 A.D., had been in common use in China for four centuries before the Gutenberg Bible appeared in Europe in 1448 A.D. Besides, the Chinese script being pictographic, with as many as forty-thousand different characters, several thousands of which in common use, type-setting was a much more onerous task for Bi Sheng than it was for Gutenberg dealing with an alphabetic script.

Even in the centuries which followed, although China had also, once again, to suffer "barbarian invasions" in its turn, the lead was maintained for a while in certain areas. For example, China was at one stage for several hundred years the world's dominant sea-power. In the early fifteenth century, a Ming imperial fleet comprising sixty ocean-going vessels visited the East African coast, some decades before Columbus set sail for America. Only after the industrial revolution unique to Europe, which made traditional methods of production obsolete, had China definitely fallen behind. A very brief outllne of Chinese history of this latter period not covered by our main narrative is given in the Roundoff section of the text.

Appendix II: Historical Sources, Original Poems

The historical narrative presented in the text is little more than an impressionistic outline peppered with occasional details for illustration, and has of course no merit in scholarly accuracy or in conceptual originality. It is just an impression that I have gathered for myself from a lifetime of amateurish, perhaps even casual, reading of the historical classics such as the *Zuozhuan, Shiji, Hanshu,* and *Zizhi Tongjian,* supplemented by some modern treatises such as the *Zhongguo Shigao* by Guo Moruo et al. and the *Zhongguo Tongshi* by Fan Wenlan et al., plus some specialized histories of certain periods or on certain topics, such as the *Zhanguo Shi* on the Warring States period by Yang Kuan. Where the classics have been defective in concentrating only on the mainly Han-ethnic Middle Kingdom culture, I have been greatly helped in my understanding of the roles played by other ethnic groups by a recent series on the subject published by the Guangxi Normal University Press, especially the volumes on the Di, Xiongnu and related nations by Ma Changshou, Duan Lianqin, and Yang Shengmin, and by a collection called *Xiyu Tongshi* of essays by various authors edited by Yu Taishan and published by the Zhongzhou Guji Publishing Society. Another area not covered by the classics is, of course, prehistory as clarified by modern archaeological research, in the appreciation of which I have benefitted from various works, especially in the social aspects from a collection of essays on the beginning of agriculture and the origin of civilization by Yan Wenming published by the Chinese Academy of Social Sciences, and in the ethnological aspects from the volumes already cited.

Although most of the poems translated in this volume can be found in easily available anthologies, such as the *Tangshi Sanbaishou* for Tang poets, not all fall into this category, and even the well-known ones can exist in different versions. Following therefore the near-unanimous suggestion of my test-readers, I have included below the original texts in full.

《詩經・七月》

七月流火。九月授衣。一之日觱發。二之日栗烈。無衣無褐。何以卒歲。三之日於耜。四之日舉趾。同我婦子。饁彼南畝。田畯至喜。七月流火。九月授衣。春日載陽。有鳴倉庚。女執懿筐。遵彼微行。爰求柔桑。春日遲遲。采蘩祁祁。女心傷悲。殆及公子同歸。七月流火。八月萑葦。蠶月條桑。取彼斧斨。以伐遠揚。猗彼女桑。七月鳴鵙。八月載績。載玄載黃。我朱孔陽。為公子裳。四月秀葽。五月鳴蜩。八月其穫。十月隕蘀。一之日於貉。取彼狐貍。為公子裘。二之日其同。載纘武功。言私其豵。獻豣於公。五月斯螽動股。六月莎雞振羽。七月在野。八月在宇。九月在戶。十月蟋蟀入我牀下。穹窒熏鼠。塞向墐戶。嗟我婦子。曰為改歲。入此室處。六月食鬱及薁。七月亨葵及菽。八月剝棗。十月穫稻。為此春酒。以介眉壽。七月食瓜。八月斷壺。九月叔苴。采荼薪樗。食我農夫。九月築場圃。十月納禾稼。黍稷重穋。禾麻菽麥。嗟我農夫。我稼既同。上入執宮功。晝爾於茅。宵爾索綯。亟其乘屋。其始播百穀。二之日鑿冰沖沖。三之日納于凌陰。四之日其蚤。獻羔祭韭。九月肅霜。十月滌場。朋酒斯饗。曰殺羔羊。躋彼公堂。稱彼兕觥。萬壽無疆。

《詩經・野有死麕》

野有死麕。白茅包之。有女懷春。吉士誘之。林有樸樕。野有死鹿。白茅純束。有女如玉。舒而脫脫兮。無感我帨兮。無使尨也吠。

《詩經・擊鼓》

擊鼓其鏜。踊躍用兵。土國城漕。我獨南行。從孫子仲。平陳與宋。不我以歸。憂心有忡。爰居爰處。爰喪其馬。于以求之。于林之下。死生契闊。與子成說。執子之手。與子偕老。于嗟闊兮。不我活兮。于嗟洵兮。不我信兮。

《詩經・伯兮》

伯兮朅兮。邦之桀兮。伯也執殳。為王前驅。自伯之東。首如飛蓬。豈無膏沐。誰適為容。其雨其雨。杲杲出日。願言思伯。甘心首疾。焉得諼草。言樹之背。願言思伯，使我心痗。

《詩經・褰裳》

子惠思我。褰裳涉溱。子不我思。豈無他人。狂童之狂也且。
子惠思我。褰裳涉洧。子不我思。豈無他士。狂童之狂也且。

《詩經・綢繆》

綢繆束薪。三星在天。今夕何夕。見此良人。子兮子兮。如此良人何。
綢繆束芻。三星在隅。今夕何夕。見此邂逅。子兮子兮。如此邂逅何。
綢繆束楚。三星在戶。今夕何夕。見此粲者。子兮子兮。如此粲者何。

《詩經・氓》

氓之蚩蚩。抱布貿絲。匪來貿絲。來即我謀。送子涉淇。至於頓丘。匪我愆期。子無良媒。將子無怒。秋以為期。乘彼垝垣。以望復關。不見復關。泣涕漣漣。既見復關。載笑載言。爾卜爾筮。體無咎言。以爾車來。以我賄遷。桑之未落。其葉沃若。于嗟鳩兮。無食桑葚。于嗟女兮。無與士耽。士之耽兮。猶可說也。女之耽兮。不可說也。桑之落矣。其黃而隕。自我徂爾。三歲食貧。淇水湯湯。漸車帷裳。女也不爽。士貳其行。士也罔極。二三其德。三歲為婦。靡室勞矣。夙興夜寐。靡有朝矣。言既遂矣。至於暴矣。兄弟不知。咥其笑矣。靜言思之。躬自悼矣。及爾偕老。老使我怨。淇則有岸。隰則有泮。總角之宴。言笑晏晏。信誓旦旦。不思其反。反是不思。亦已焉哉。

屈原《天問》節選

曰遂古之初。誰傳導之。上下未形。何由考之。冥昭瞢暗。誰能極之。馮翼惟像。何以識之。明明闇闇。惟時何為。陰陽三合。何本何化。圜則九重。孰營度之。惟茲何功。孰初作之。……

屈原《河伯》

與女遊兮九河。衝風起兮水揚波。與女沐兮咸池。晞女髮兮陽之阿。望美人兮未來。臨風怳兮浩歌。乘水車兮荷蓋。駕兩龍兮驂螭。登崑崙兮四望。心飛揚兮浩蕩。日將暮兮悵忘歸。惟極浦兮寤懷。魚鱗屋兮龍堂。紫貝闕兮珠宮。靈何為兮水中。乘白黿兮逐文魚。與女遊兮河之渚。流澌紛兮將來下。子交手兮東行。送美人兮南浦。波滔滔兮來迎。魚鱗鱗兮媵予。

屈原《哀郢》

皇天之不純命兮。何百姓之震愆。民離散而相失兮。方仲春而東遷。去故都而就遠兮。遵江夏以流亡。出國門而軫懷兮。甲之鼂吾以行。發郢都而去閭兮。怊荒忽其焉極。楫齊揚以容與兮。哀見君而不再得。望長楸而太息兮。涕淫淫其若霰。過夏首而西浮兮。顧龍門而不見。心嬋媛而傷懷兮。眇不知其所蹠。順風波以從流兮。焉洋洋而為客。淩陽侯之氾濫兮。忽翱翔之焉薄。心絓結而不解兮。思蹇產而不釋。將運舟而下浮兮。上洞庭而下江。去終古之所居兮。今逍遙而來東。羌靈魂之欲歸兮。何須臾而忘反。背夏浦而西思兮。哀故都之日遠。登大墳以遠望兮。聊以舒吾憂心。哀州土之平樂兮。悲江介之遺風。當陵陽之焉至兮。淼南渡之焉如。曾不知夏之為丘兮。孰兩東門之可蕪。心不怡之長久兮。憂與愁其相接。惟郢路之遼遠兮。江與夏之不可涉。忽若去不信兮。至今九年而不復。慘鬱鬱而不通兮。蹇侘傺而含慼。外承歡之汋約兮。諶荏弱而難持。忠湛湛而願進兮。妒被離而鄣之。堯舜之抗行兮。瞭杳杳而薄天。眾讒人之嫉妒兮。被以不慈之偽名。憎慍惀之脩美兮。好夫人之忼慨。眾踥蹀而日進兮。美超遠而逾邁。亂曰。曼余目以流觀兮。冀壹反之何時。鳥飛反故鄉兮。狐死必首丘。信非吾罪而棄逐兮。何日夜而忘之。

項羽《垓下歌》

力拔山兮氣蓋世。時不利兮騅不逝。騅不逝兮可奈何。虞兮虞兮奈若何。

無名氏《悲歌》

悲歌可以當泣。遠望可以當歸。思念故鄉。鬱鬱纍纍。
欲歸家無人。欲渡河無船。心思不能言。腸中車輪轉。

無名氏《別詩》

結髮為夫妻。恩愛兩不疑。歡娛在今夕。嬿婉及良時。征夫懷往路。起視夜何其。參辰皆已沒。去去從此辭。行役在戰場。相見未有期。握手一長歎。淚為生別滋。努力愛春華。莫忘歡樂時。生當復來歸。死當長相思。

無名氏《戰城南》

戰城南。死郭北。野死不葬烏可食。為我謂烏。且為客豪。野死諒不葬。腐肉安能去子逃。水深激激。蒲葦冥冥。梟騎戰鬥死。駑馬徘徊鳴。梁築室。何以南。何以北。禾黍不獲君何食。願為忠臣安可得。思子良臣。良臣誠可思。朝行出攻。暮不夜歸。

無名氏《十五從軍征》

十五從軍征。八十始得歸。道逢鄉里人。家中有阿誰。遙看是君家。松柏塚累累。兔從狗竇入。雉從梁上飛。中庭生旅穀。井上生旅葵。舂穀持作飯。採葵持作羹。羹飯一時熟。不知貽阿誰。出門東向看。淚落沾我衣。

無名氏《猛虎行》

饑不從猛虎食。莫不從野雀棲。野雀安無巢。遊子為誰驕。

無名氏《上邪》

上邪。我欲與君相知。長命無絕衰。山無陵。江水為竭。冬雷震震。夏雨雪。天地合。乃敢與君绝。

無名氏《客從遠方來》

客從遠方來。遺我一端綺。相去萬餘里。故人心尚爾。文彩雙鴛鴦。裁為合歡被。著以長相思。緣以結不解。以膠投漆中。誰能別離此。

陳琳《飲馬長城窟行》

飲馬長城窟。水寒傷馬骨。往謂長城吏。慎莫稽留太原卒。官作自有程。舉築諧汝聲。男兒寧當格鬬死。何能怫鬱築長城。長城何連連。連連三千里。邊城多健少。內舍多寡婦。作書與內舍。便嫁莫留住。善侍新姑嫜。時時念我故夫子。報書往邊地。君今出語一何鄙。身在禍難中。何為稽留他家子。生男慎莫舉。生女哺用脯。君獨不見長城下。死人骸骨相撐拄。結髮行事君。慊慊心意關。明知邊地苦。賤妾何能久自全。

蔡琰《胡笳十八拍》節選

……

為天有眼兮何不見我獨漂流。為神有靈兮何事處我天南海北頭。我不負天兮天何配我殊匹。我不負神兮神何殛我越荒州。製兹八拍兮擬俳優。何知曲成兮心轉愁。……

城頭烽火不曾滅。疆場征戰何時歇。殺氣朝朝衝塞門。胡風夜夜吹邊月。故鄉隔兮音塵絕。哭無聲兮氣將咽。一生辛苦兮緣別離。十拍悲深兮淚成血。……

我非貪生而惡死。不能捐身兮心有以。生仍冀得兮歸桑梓。死當埋骨兮長已矣。日居月諸兮在戎壘。胡人寵我兮有二子。鞠之育之兮不羞恥。愍之念之兮生長邊鄙。十有一拍兮因茲起。哀響纏綿兮徹心髓。……

東風應律兮暖氣多。知是漢家天子兮布陽和。羌胡蹈舞兮共謳歌。兩國交歡兮罷兵戈。忽遇漢使兮稱近詔。遺千金兮贖妾身。喜得生還兮逢聖君。嗟別稚子兮會無因。十有二拍兮哀樂均。去住兩情兮難具陳。……

十六拍兮思茫茫。我與兒兮各一方。日東月西兮徒相望。不得相隨兮空斷腸。對萱草兮憂不忘。彈鳴琴兮情何傷。今別子兮歸故鄉。舊怨平兮新怨長。泣血仰頭兮訴蒼蒼。胡為生我兮獨罹此殃。……

曹操《短歌行》

對酒當歌。人生幾何。譬如朝露。去日苦多。慨當以慷。幽思難忘。何以解憂。唯有杜康。青青子衿。悠悠我心。但為君故。沈吟至今。呦呦鹿鳴。食野之苹。我有嘉賓。鼓瑟吹笙。明明如月。何時可掇。憂從中來。不可斷絕。越陌度阡。枉用相存。契闊談讌。心念舊恩。月明星稀。烏鵲南飛。繞樹三匝。何枝可依。山不厭高。海不厭深。周公吐哺。天下歸心。

曹植《七步歌》

煮豆燃豆萁。豆在釜中泣。本是同根生。相煎何太急。

無名氏《隴頭歌辭》

隴頭流水。流離山下。念吾一身，飄然曠野。
朝發欣城。暮宿隴頭。寒不能語。舌捲入喉。
隴頭流水。鳴聲嗚咽。遙望秦川。心肝斷絕。

無名氏《孟津河》

遙看孟津河。楊柳鬱婆娑。我是虜家兒。不解漢兒歌。

無名氏《門前一株棗》

門前一株棗。歲歲不知老。阿婆不嫁女。那得孫兒抱。

無名氏《華陰山頭井》

華陰山頭百丈井。下有流水徹骨冷。可憐女子能照影。不見其餘見斜領。

無名氏《三峽謠》

朝發黃牛。暮宿黃牛。三朝三暮。黃牛如故。

陶淵明《結廬在人境》

結廬在人境。而無車馬喧。問君何能爾。心遠地自偏。採菊東籬下。悠然見南山。山氣日夕佳。飛鳥相與還。此中有真意。欲辯已忘言。

鮑照《對案不能食》

對案不能食。拔劍擊柱長歎息。丈夫生世會幾時。安能蹀躞垂羽翼。棄置罷官去。還家自休息。朝出與親辭。暮還在親側。弄兒床前戲。看婦機中織。自古聖賢盡貧賤。何況我輩孤且直。

庾信《家住金陵縣前》

家住金陵縣前。嫁得長安少年。回頭望鄉淚落。不知何處天邊。
胡塵幾日應盡。漢月何時更圓。為君能歌此曲。不覺心隨斷弦。

陳子昂《登幽州臺歌》

前不見古人。後不見來者。念天地之悠悠。獨愴然而涕下。

李隆基《經魯祭孔子》

夫子何為者。棲棲一代中。地猶鄹氏邑。宅即魯王宮。
歎鳳嗟身否。傷麟怨道窮。今看兩楹奠。當與夢時同。

杜甫《望嶽》

岱宗夫如何。齊魯青未了。造化鍾神秀。陰陽割昏曉。
蕩胸生層雲。決眥入歸鳥。會當凌絕頂。一覽眾山小。

李白《早發白帝城》

朝辭白帝彩雲間。千里江陵一日還。兩岸猿聲啼不住。輕舟已過萬重山。

王維《渭川田家》

斜陽照墟落。窮巷牛羊歸。野老念牧童。倚杖候荊扉。雉雊麥苗秀。
蠶眠桑葉稀。田夫荷鋤至。相見語依依。即此羨閒逸。悵然吟式微。

崔顥《長干行》

君家何處住。妾住在橫塘。停船暫借問。或恐是同鄉。
家臨九江水。來去九江側。同是長干人。生小不相識。

王維《送別》

下馬飲君酒。問君何所之。君言不得意。
歸臥南山陲。但去莫復問。白雲無盡時。

李白《將進酒》

君不見黃河之水天上來。奔流到海不復回。君不見高堂明鏡悲白髮。朝如青絲暮成雪。人生得意須盡歡。莫使金樽空對月。天生我材必有用。千金散盡還復來。烹羊宰牛且為樂。會須一飲三百杯。岑夫子。丹丘生。將進酒。杯莫停。與君歌一曲。請君為我傾耳聽。鐘鼓饌玉不足貴。但願長醉不願醒。古來聖賢皆寂寞。唯有飲者留其名。陳王昔時宴平樂。斗酒十千恣讙謔。主人為何言少錢。徑須沽取對君酌。五花馬。千金裘。呼兒將出換美酒。與爾同銷萬古愁。

李白《月下獨酌》

花間一壺酒。獨酌無相親。舉杯邀明月。對影成三人。月既不解飲。影徒隨我身。暫伴月將影。行樂須及春。我歌月徘徊。我舞影零亂。醒時同交歡。醉後各分散。永結無情遊。相期邈雲漢。

杜甫《兵車行》

車轔轔。馬蕭蕭。行人弓箭各在腰。爺娘妻子走相送。塵埃不見咸陽橋。牽衣頓足攔道哭。哭聲直上干雲霄。道旁過者問行人。行人但云點行頻。或從十五北防河。便至四十西營田。去時里正與裹頭。歸來頭白還戍邊。邊庭流血成海水。武皇開邊意未已。君不聞漢家山東。二百州。千村萬落生荊杞。縱有健婦把鋤犁。禾生隴畝無東西。況復秦兵耐苦戰。被驅不異犬與雞。長者雖有問。役夫敢申恨。且如今年冬。未休關西卒。縣官急索租。租稅從何出。信知生男惡。反是生女好。生女猶得嫁比鄰。生男埋沒隨百草。君不見。青海頭。古來白骨無人收。新鬼煩冤舊鬼哭。天陰雨濕聲啾啾。

李白《胡無人》

嚴風吹霜海草凋。筋幹精堅胡馬驕。漢家戰士三十萬。將軍兼領霍嫖姚。流星白羽腰間插。劍花秋蓮光出匣。天兵照雪下玉關。虜箭如沙射金甲。雲龍風虎盡交回。太白入月敵可摧。敵可摧。旄頭滅。履胡之腸涉胡血。懸胡青天上。埋胡紫塞旁。胡無人。漢道昌。陛下之壽三千霜。但歌大風雲飛揚。安用猛士兮守四方。

杜甫《月夜憶舍弟》

戍鼓斷人行。秋邊一雁聲。露從今夜白。月是故鄉明。
有弟皆分散。無家問死生。寄書長不達。況乃未休兵。

李益《喜見外弟又言別》

十年離亂後。長大一相逢。問姓驚初見。稱名憶舊容。
別來滄海事。語罷暮天鐘。明日巴陵道。秋山又幾重。

黃幡綽《嘲劉文樹》

可憐好個劉文樹。髭鬚共頦頤別住。文樹面孔不似猢猻。猢猻面孔強似文樹。

杜秋娘《金縷衣》

勸君莫惜金縷衣。勸君惜取少年時。花開堪折直須折。莫待無花空折枝。

金昌緒《春怨》

打起黃鶯兒。莫教枝上啼。啼時驚妾夢。不得到遼西。

坎曼爾《訴豺狼》

東家豺狼惡。食吾饢。飲吾血。五穀未離場。大布未下機。已非吾所有。
有朝一日。天崩地裂豺狼死。吾卻雲開復見天。

杜牧《過驪山作》

始皇東遊出周鼎。劉項縱觀皆引頸。削平天下實辛勤。卻為道旁窮百姓。
黔首不愚爾益愚。千里函關囚獨夫。牧童火入九泉底。燒作灰時猶未枯。

李賀《金銅仙人辭漢歌》

茂陵劉郎秋風客。夜聞馬嘶曉無跡。畫欄桂樹懸秋香。三十六宮土花碧。
魏官牽車指千里。東關酸風射眸子。空將漢月出宮門。憶君清淚如鉛水。
衰蘭送客咸陽道。天若有情天亦老。攜盤獨出月荒涼。渭城已遠波聲小。

王維《桃源行》

漁舟逐水愛山春。兩岸桃花夾古津。坐看紅樹不知遠。行盡青溪忽值人。
山口潛行始隈隩。山開曠望旋平陸。遙看一處攢雲樹。近入千家散花竹。
樵客初傳漢姓名。居人未改秦衣服。居人共住武陵源。還從物外起田園。
月明松下房櫳靜。日出雲中雞犬喧。驚聞俗客爭來集。競引還家問都邑。
平明閭巷掃花開。薄暮漁樵乘水入。初因避地去人間。更問神仙遂不還。
峽裡誰知有人事。世中遙望空雲山。不疑靈境難聞見。塵心未盡思鄉縣。
出洞無論隔山水。辭家終擬長游衍。自謂經過舊不迷。安知峰壑今來變。
當時只記入山深。青溪幾度到雲林。春來遍是桃花水。不辨仙源何處尋。

白居易《花非花》

花非花。霧非霧。夜半來。天明去。來如春夢不多時。去似朝雲無覓處。

孟郊《遊子吟》

慈母手中線。遊子身上衣。臨行密密縫。意恐遲遲歸。誰言寸草心。報得三春暉。

白居易《問劉十九》

綠螘新醅酒。紅泥小火爐。晚來天欲雪。能飲一杯無。

李商隱《夜雨寄北》

君問歸期未有期。巴山夜雨漲秋池。何當共剪西窗燭。卻話巴山夜雨時。

劉禹錫《水》

水。至清。盡美。從一勺。至千里。利人利物。時行時止。道性淨皆然。交情淡如此。君游金谷堤上。我在石渠署裏。兩心相憶似流波。潺湲日夜無窮已。

李商隱《登樂遊原》

向晚意不適。驅車登古原。夕陽無限好。只是近黃昏。

李白《山中問答》

問余何意棲碧山。笑而不答心自閑。桃花流水窅然去。別有天地非人間。

李商隱《無題》

相見時難別亦難。東風無力百花殘。春蠶到死絲方盡。蠟炬成灰淚始乾。曉鏡但愁雲鬢改。夜吟應覺月光寒。蓬萊此去無多路。青鳥殷勤為探看。

杜牧《贈別》

多情卻似總無情。唯覺尊前笑不成。蠟燭有心還惜別。替人垂淚到天明。

黃巢《不第後賦菊》

待到秋來九月八。我花開後千花殺。沖天香陣透長安。滿城盡帶黃金甲。

韋莊《菩薩蠻》

人人盡說江南好。遊人只合江南老。春水碧於天。畫船聽雨眠。鑪邊人似月。皓腕凝霜雪。未老莫還鄉。還鄉須斷腸。

李煜《相見歡》

無言獨上西樓。月如鉤。寂寞梧桐深院鎖清秋。剪不斷。理還亂。是離愁。別是一般滋味在心頭。

李煜《 相見歡》

林花謝了春紅。太忽忽。無奈朝來寒雨晚來風。胭脂淚。相和醉。幾時重。自是人生長恨水長東。

李煜《虞美人》

春花秋月何時了。往事知多少。小樓昨夜又東風。故國不堪回首月明中。雕欄玉砌應猶在。只是朱顏改。問君還有幾多愁。恰似一江春水向東流。

王安國《清平樂》

留春不住。費盡鶯兒語。滿地殘紅宮錦污。昨夜南園風雨。小憐初上琵琶。曉來思繞天涯。不肯畫堂朱戶。春風自在楊花。

黃庭堅《清平樂》

春歸何處。寂寞無行路。若有人知春去處。喚取歸來同住。春無蹤迹誰知。除非問取黃鸝。百囀無人能解。因風飛過薔薇。

蘇軾《和子由澠池懷舊》

人生到處知何似。應似飛鴻踏雪泥。泥上偶然留指爪。鴻飛那復計東西。老僧已死成新塔。壞壁無由見舊題。往日崎嶇還記否。路長人困蹇驢嘶。

蘇軾《梅花》

何人把酒慰深幽。開自無聊落更愁。幸有清溪三百曲。不辭相送到黃州。

蘇軾《書雙竹湛師房》

暮鼓朝鐘自擊撞。閉門孤枕對殘釭。白灰旋撥通紅火。臥聽蕭蕭雨打窗。

蘇軾《卜算子》

水是眼波橫。山是眉峰聚。欲問行人去那邊。眉眼盈盈處。
才始送春歸。又送君歸去。若到江南趕上春。千萬和春住。

范仲淹《漁家傲》

塞下秋來風景異。衡陽雁去無留意。四面邊聲連角起。千嶂裏。長煙落日孤城閉。濁酒一杯家萬里。燕然未勒歸無計。羌管悠悠霜滿地。人不寐。將軍白髮征夫淚。

王安石《促織》

金屏翠幔與秋宜。得此年年醉不知。只向貧家促機杼。幾家能有一絢絲。

王安石《白頭想見江南》

柳葉鳴蜩綠暗。荷花落日紅酣。三十六陂春水。白頭想見江南。

蘇軾《秋早川原淨麗》

秋早川原淨麗。雨餘風日清酣。從此歸耕劍外。何人送我池南。

蔣興祖女《減字木蘭花》

朝雲橫渡。轆轆車聲如水去。白草黃沙。月照孤村三兩家。
飛鴻過也。萬結愁腸無晝夜。漸近燕山。回首鄉關歸路難。

李清照《永遇樂》

落日鎔金。暮雲合璧。人在何處。染柳煙濃。吹梅笛怨。春意知幾許。元宵佳節。融和天氣。次第豈無風雨。來相召。香車寶馬。謝他酒朋詩侶。中州盛日。閨門多暇。記得偏重三五。鋪翠冠兒。撚金雪柳。簇帶爭濟楚。如今憔悴。風鬟霧鬢。怕見夜間出去。不如向簾兒底下。聽人笑語。

岳飛《滿江紅》

遙望中原。荒煙外。許多城郭。想當年。花遮柳護。鳳樓龍閣。萬歲山前珠翠繞。蓬壺殿裡笙歌作。到而今。鐵騎滿郊畿。風塵惡。兵安在。膏鋒鍔。民安在。填溝壑。歎江山如故。千村寥落。何日請纓提鋭旅。一鞭直渡清河洛。卻歸來。重續漢陽遊。騎黃鶴。

辛棄疾《采桑子》

少年不識愁滋味。愛上層樓。愛上層樓。為賦新詞強說愁。
而今識盡愁滋味。欲說還休。欲說還休。卻道天涼好個秋。

《義勇軍進行曲》（中華人民共和國國歌）田漢 詞　　聶耳 曲

起來！不願做奴隸的人們！把我們的血肉，築成我們新的長城！中華民族到了最危險的時候，每個人被迫着發出最後的吼聲，起來！起來！起來！我們萬眾一心，冒着敵人的炮火前進，冒着敵人的炮火前進！前進！前進！進！

《歷史的傷口》佚名

矇上眼睛。就以為看不見。掩上耳朵。就以為聽不到。而真理在心中。創痛在胸口。還要忍多久。還要沉默多久。如果熱淚可以洗淨塵埃。如果熱血可以換來自由。讓明天能記得今天的怒吼。讓世界都看到歷史的傷口。

游順釗《墨淚》

留下逃亡的足跡。口號嘶啞的回響。平添的手鐐。妻女的無靠。還沒冷卻的熱血。再不跳動的脈搏。而遠離中土的我。就僅僅紙上這幾點墨。

淪西志士《匹夫之志》

僅負興亡責。因名莫折腰。英雄灰土矣。豈待看今朝。

Appendix III: A Recital of the Poems

As emphsized several times already in the text, a very important property of a classical Chinese poem is its tonal quality, which makes it sound almost like a song when read aloud, and this tonal quality is, of course, lost in English translation. To give my readers an idea how an ordinary Chinese lover of poetry would read a classical poem for his/her own enjoyment, I have made available on the web, by curtesy of World Scientific, my publisher, a recorded recital by myself of the poems included in this volume, that is, all but three near the end not in the classical style.

Some words of warning, however, would be appropriate. First, I am not at all an expert performer in recitation, nor do I have a good voice, but this need not be too important, for most of classical Chinese poetry was probably intended for private appreciation, not for performance. And even for those genres meant to be performed, such as the *yuefu* and the *ci*, expertise may not be of much help as regards authenticity, since it is not known now exactly how they were performed. Secondly, I should add that my recital is given in the Cantonese dialect, my mother tongue, not in Putong Hua (Mandarin), at present the national dialect. Cantonese is believed, at least by some, to approximate the classical dialect in which these poems were originally written, probably better than Putong Hua does. For instance, the classical dialect was said to have eight tones (sheng), while Cantonese has nine, but Putong Hua has in general only four. Practically, what matters is the fact that poetry from before the end of the Northern Song dynasty (12th century A.D.), the period from which almost all the poems in this volume are selected, sounds good in Cantonese, but the tonal quality starts to decline in Cantonese for poetry from later periods. This can be taken as an indication that Chinese dialects had much evolved since then as a result of the massive influx of non-Han ethnic groups from the north. In that case, the northern dialect, on which Putong Hua is based, would have changed more than the Cantonese dialect in the south. In any case, reciting in Cantonese is for me unavoidable, for my Putong Hua pronunciation is unfortunately not good enough for reciting poetry.

Website: http://www.worldscibooks.com/eastasianstudies/8031.html

Appendix IV: Captions of Illustrations

The illustrations were originally meant merely for decoration and for filling the blank pages left over by those poems covering more than one page. They were selected just from my own small collection of rubbings and from books in my possession, the criteria being only that I liked them and that they reproduced well. But once included, they seem to have acquired a significance of their own, not only to myself but also to some of my test-readers who are interested enough to suggest that explanations be given of what they represent, thinking that these may show another facet of life in the period described by the poems and the narrative. So here they are.

Dedication page: Seals carved for my father in his lifetime by my brother Hong-Fat (Henry). The pair on the extreme right and left says, loosely: "(The sage) puts himself behind, hence ends up in front; has scant regard for himself, hence is left with a greater Self," which is a quotation from Laozi. The second from the right is also a quotation from Laozi and it says: "The highest virtue is like water (which benefits all living things but strives not for itself, seeking only the low position that others hate)". The second from the left is, I think, Confucian, and it says: "Be not biassed towards some because they share with you similarities, and repudiate not others just because they differ from you." The other three are just personal seals of us brothers who share in the dedication.

Last page of Contents: Another seal from the same collection with the legend: "Wild-goose prints on snow-bound earth", referring to a famous poem by Su Shi found on page 179.

Page 7 and throughout the volume, Chinese titles of sections and of poems by my brother Hong-Ching (Eric) in his special calligraphic style.

Page 15: Characters (in an ancient script) engraved on the inside of a bronze vessel from Western Zhou about the time when the poem "*Seventh Month*" was composed. They actually make up the last line of that poem and mean: "Ten thousand years without limit", which are still used in modern Chinese as a rough equivalent to the birthday wish: "Many happy returns" in English. Reproduced from a publication of the National Palace Museum (故宮博物院), Taipei, Taiwan, (2001), entitled: "Expression in Bronze: Ancient Inscription of the Western Chou (千古金言話西周)", by kind permission of the author Tu (Zhengsheng) (杜正勝).

Page 37: A weapon in bronze of the Warring States period (?) of the type known as *ge* which is basically a dagger mounted at the end of a long handle. Reproduced from a publication of the Shanghai Guji Publishing Company (上海古籍出版社) (1990), entitled: "Rubbings of Ancient Weapons from the Zungu Zhai Collection (尊古齋古兵精拓)", edited by Huang Jun (黄濬).

Page 65: Ceramic tile of the type known as *wadang*. Tiles of this type were attached to the ends of rafters on tiled roofs for decoration and for protection of the rafters from the weather. Han dynasty is famous for its *wadang*, but I am not sure this one is from the Han dynasty. Reproduced from a publication of the Shanghai Guji Publishing Company (上海古籍出版社)(1990), entitled: "Inscriptions on *Wadang* from the Zungu Zhai Collection (尊古齋瓦當文字)", edited by Huang Jun (黄濬).

Page 73: Bas-relief on a Han dynasty brick. On such bricks vivid scenes of contemporary life are often depicted. Reproduced from a rubbing in the author's possession.

Page 99: A Tang dynasty tile found on the Silk Route. Reproduced from a rubbing in the author's possession. Notice the distinctly non-Han ethnic features of the person depicted.

Page 115: Porcelain jug from the golden age of Li Longji's reign. Reproduced from a photograph in a publication of the National Museum of China, Beijing,

(1998), entitled: "Historical Exhibits in the Chinese Historical Museum (中國歷史博物館中國通史陳列)", by kind permission of the museum (general office).

Page 121: One of four figures mounted on a camel of glazed porcelain in typical Tang three-coloured style. Reproduced from a photograph in a publication of the National Museum of China, Beijing, (1998), entitled: "Historical Exhibits in the Chinese Historical Museum (中國歷史博物館中國通史陳列", by kind permission of the museum (general office). Notice again the non-Han ethnic features of the person depicted.

Page 143: Painting reputedly by Wang Wei depicting a snowy scene. Reproduced from a photograph in the book entitled "Chinese Painting" by Peter C. Swann, Éditions Pierre Tisné, Paris, (1958).

Page 187: Song dynasty painting (ink on paper) of six persimmons by Mu Xi (Mu Ch'i). Reproduced from a photograph in the book entitled "Chinese Painting" by Peter C. Swann, Éditions Pierre Tisné, Paris, (1958).

Appendix V: Glossary, Chinese Names and Terms

There being often tens of words corresponding to the same sound in the Chinese language, even when tones are taken into account, it can be difficult to identify the actual characters meant, given just their pinyin. This appendix is intended to help the readers who know Chinese to make the appropriate correspondence.

A

Aiying 哀郢
Almberg-Ng, Evangeline 吳兆朋

B

Bai Juyi 白居易
Baidi 白帝
Ban Chao 班超
Bao Zhao 鮑照
Bashan 巴山
beacon fires 烽火
Beijing 北京
Bi Sheng 畢昇
Brown Ox (Hill) 黃牛

C

Cai Lun 蔡倫
Cai Yan 蔡琰
Cai Yong 蔡邕
Cao Cao 曹操
Cao Pi 曹丕
Cao Zhi 曹植
Cen (Sir) 岑（夫子）
Central Plains 中原
Chan Hong Ching (Eric) 陳匡正
Chan Hong Fat (Henry) 陳匡法
Chan Hong-Mo 陳匡武
Chan, Man-Kwun 陳曼琨
Chan, Man-Suen 陳曼珣
Chang'an 長安
Changgan 長干
Changjiang 長江
Chen 陳
Chen Lin 陳琳
Chen Sheng (She) 陳勝（涉）
Chen Zi'ang 陳子昂
Chibi (Battle of) 赤壁（之戰）
Chief of Kings 霸王
Chongqing 重慶
Chu 楚
Chunqiu 春秋
ci 詞
cipai 詞牌
Confucius (Kongzi) 孔子
couplets 對聯

Cui Hao 崔顥
cuzhi 促織

D

Dan Sheng 丹丘生
di 帝
Di (Dili) 狄、翟（丁零、狄歷）
Donghu 東胡
Dongting (Lake) 洞庭（湖）
Du Fu 杜甫
Du Mu 杜牧
Du Qiuniang 杜秋娘
Duan Lianqin 段連勤
Dunhuang 敦煌
Dunqi (Hill) 頓丘

E

Early Tang 初唐
Eastern Han 東漢
Eastern Jin 東晉
Eastern Zhou 東周

F

Fa Xian 法顯
Fan Wenlan 范文瀾
Fan Zhongyan 范仲淹
Feishui (Battle of) 淝水（之戰）
Festival of Lights 元宵
Five Dynasties 五代
Fu Fei 宓妃
Fu Jian 苻堅

G

ganlu 甘露
Gansu 甘肅
Gaozong 高宗
Gaozu 高祖
ge 戈
Genghis Khan 成吉思汗
Gebi (Gobi) Desert 戈壁（沙漠）
Great Khan from Heaven 天可汗
Guan Zhong 管仲
Guangxi Normal 廣西師範
Guo Moruo 郭沫若
gushi 古詩

H

Hainan (Island) 海南島
Han (dynasty) 漢朝
Han (ethnic group) 漢（族）
Han (River) 漢（江）
Han (Warring State) 韓（國）
Hangzhou 杭州
Hanshu 漢書
He Bo 河伯
Henan 河南
Hengtang 横塘
heshi 和詩
Hu 胡
Huang Chao 黃巢
Huang Fanchuo 黃幡綽
Huang Tingjian 黃庭堅
Huanghe 黃河
Huangzhou 黃州
Huayin 華陰
Huihe 回紇
hujia 胡笳
Hunan 湖南
Huo (Commander) 霍（嫖姚）

J

Jia (day) 甲（日）
Jiang 江
Jiang Xingzu 蔣興祖
Jiangling 江陵
Jiangnan 江南
Jiangsu 江蘇
Jiankang (Jianye) 建康（建業）
Jin (265–420 A.D.) 晉
Jin (1112–1234 A.D.) 金
Jin Changxu 金昌緒
Jingdi 景帝
Jinling 金陵
Jiuge 九歌
Jiujiang 九江
Jiulishan 九里山
jueju 絕句
jujube 棗

K

Kaifeng 開封
Kaiyuan 開元
Kan Man Er 坎曼爾
Kashgar 喀什噶爾（疏勒）
Kongzi (Confucius) 孔子
Kunlun 崑崙

L

Laozi (Lao Tzu) 老子
Late Tang 晚唐
Latter Wei 後魏
Legalist (School) 法家
li (1/2 Km) 里
Li Bai (Li Bo, Li Po) 李白
Li He 李賀
Li Longji 李隆基
Li Qingzhao 李清照
Li Shangyin 李商隱
Li Shimin 李世民
Li Yi 李益
Li Yu (Houzhu) 李煜（後主）
Li Yuan 李淵
Li Zhongzhu 李中主
Liang 梁
Liangzhu (culture) 良渚（文化）
Liao 遼
Lin'an 臨安
Linzi 臨淄
Lisao 離騷
Liu Bang 劉邦
Liu Wenshu 劉文樹
Liu Xiu 劉秀
Liu Yuan 劉淵
Liu Yuxi 劉禹錫
Longtou 隴頭
Lu 魯
Lunxizhishi 淪西志士
Luo (River) 洛（水）
Luoyang 洛陽
lushi 律詩

M

Ma Changshou 馬長壽
Mao Zedong 毛澤東
Maoling 茂陵
Marquis of Wei 衛侯
Meng Jiao 孟郊
Mengjin 孟津

Middle Kingdom 華夏
Middle Tang 中唐
Milan 米蘭
Ming 明
Mu Xi (Mu Qi) 牧溪

N

Nanjing 南京
Nie Er 聶耳
Nineteen Old Poems 古詩十九首
Northern and Southern Dynasties 南北朝
Northern Song 北宋

O

Orion 三星（參星）

P

parasol (tree) 梧桐
Pengze 彭澤
ping (sheng) 平聲
Pingdi 平帝
pinyin 拼音
pipa 琵琶

Q

Qi 齊
Qi (River) 淇
Qianqin 前秦
Qin 秦
qin 琴
Qing 清
Qing Miao 青苗
Qinghai 青海
Qiuci 龜茲
qu (pai) 曲（牌）
qu (sheng) 去聲
Qu Yuan 屈原
Qufu 曲阜

R

rouge-tinted tears 胭脂淚
ru (sheng) 入聲

S

Sanxia 三峽
Sanxingdui 三星堆
Shaanxi 陝西
Shandong 山東
Shang (dynasty) 商（朝）
shang (sheng) 上聲
sheng 聲
Shenzong 神宗
shi 詩
Shi Sheng 詩聖
Shi Xian 詩仙
Shihuangdi 始皇帝
Shiji 史記
Shijing 詩經
Shu 蜀
Shun 舜
Sichuan 四川
Sima Guang 司馬光
Sima Qian 司馬遷
Song 宋
South Mount （終）南山
South Tang 南唐
Southern Song 南宋
Spring and Autumn (Period) 春秋
Su Shi (Dongpo) 蘇軾（東坡）
Su Xun 蘇洵

Su Zhe (Ziyou) 蘇轍（子游）
Sui 隋
Sun Zizhong 孫子仲
Suzhou 蘇州

T

Taishan 泰山
Taiyuan 太原
Taizong 太宗
Tan Qixiang 譚其驤
Tang 唐
Tang Golden Age 盛唐
Tangren Jie 唐人街
Tangshi Sanbaishou 唐詩三百首
Tao (Dao) (-ism) 道教
Tao Yuanming 陶淵明
terracotta army 兵馬俑
Three Gorges 三峽
Three Kingdoms 三國
tian (ci) 填（詞）
Tian Han 田漢
Tiananmen 天安門
Tianwen 天問
Todd, David 達安輝
Tsou, Judy 周尚則
Tsou Sheung Tsun 周尚真
Tufan 吐蕃

U

Urumqi 烏魯木齊
Uygur (ethnic group) 維吾爾（族）

W

wadang 瓦當
wang 王
Wang Anguo 王安國
Wang Anshi 王安石
Wang Wei 王維
Warring States 戰國
Wei 魏
Wei (River) 洧
Wei Zhuang 韋莊
Weicheng 渭城
Wendi 文帝
Western Han 西漢
Wu Di 五帝
Wu Zetian 武則天
Wudi 武帝
Wuhan 武漢
Wujiang 烏江

X

Xia (dynasty) 夏（朝）
Xia (River) 夏（江）
Xi'an 西安
Xianchi (Pool) 咸池
Xiang (River) 湘（江）
Xiang Yu 項羽
Xianyang 咸陽
Xianzong 憲宗
Xin Qiji 辛棄疾
Xincheng 欣城
Xinjiang 新疆
Xiongnu 匈奴
Xiongzhou 雄州
Xiyu Tongshi 西域通史
Xuan Zhuang 玄奘
Xuanzong 玄宗

Y

Yan 燕
Yan Wenming 嚴文明
Yan'an 延安
Yanshan 燕山
Yang 陽
Yang Guifei 楊貴妃
Yang Jian 楊堅
Yang Kuan 楊寬
Yang Shengmin 楊聖敏
Yangtze (River) 揚子（江）
Yanran 燕然
Yao 堯
Yau Shun-chiu 游順釗
Yin 陰
Ying 郢
Ying Zheng 嬴政
Youzhou 幽州
Yu (Lady) 虞（姬）
Yu Taishan 余太山
Yu Xin 庾信
Yuan (River) 沅
Yue Fei 岳飛
Yuechi 月氏
yuefu 樂府
Yumen (Pass) 玉門關
Yuwen Tai 宇文泰

Z

ze (sheng) 仄聲
Zhang Qian 張騫
Zhanguo Shi 戰國史
Zhao 趙
Zhao Mingcheng 趙明誠
Zhejiang 浙江
Zhen (River) 溱
Zheng 鄭
Zhongguo Shigao 中國史稿
Zhongguo Tongshi 中國通史
Zhongnan (Mount) 終南山
Zhongzhou Guji 中州古籍
Zhou 周
Zhougong 周公
Zizhi Tongjian 資治通鑒
zongzi 粽子
Zoushi 鄹氏
Zuozhuan 左傳